PSpice®: A Tutorial

PSpice®: A Tutorial

L. H. FENICAL

REGENTS/PRENTICE HALL
Englewood Cliffs, New Jersey 07632

Library of Congress Cataloging-in-Publication Data

Fenical, L. H.
    PSpice : a tutuorial / L. H. Fenical.
        p.   cm.
    Includes bibliographical references (p.   ) and index.
    ISBN 0-13-681149-3
    1. PSpice (Computer program)   2. Electric circuit analysis—Data
processing.   I. Title.
TK454.F46   1992   91-44413
621.319′2′0285—dc20   CIP

Editorial/production supervision and
    interior design: *Eileen M. O'Sullivan*
Cover design: Lundgren Graphics
Manufacturing buyer: *Ed O'Dougherty*
Prepress buyer: *Ilene Levy*

© 1992 by Regents Prentice Hall
A Division of Simon & Schuster
Englewood Cliffs, New Jersey 07632

PSpice® is a registered trademark of MicroSim Corporation.

Printed in the United States of America

10 9 8 7 6 5 4 3 2 1

ISBN   0-13-681149-3

PRENTICE-HALL INTERNATIONAL (UK) LIMITED, *London*
PRENTICE-HALL OF AUSTRALIA PTY. LIMITED, *Sydney*
PRENTICE-HALL CANADA INC., *Toronto*
PRENTICE-HALL HISPANOAMERICANA, S.A., *Mexico*
PRENTICE-HALL OF INDIA PRIVATE LIMITED, *New Delhi*
PRENTICE-HALL OF JAPAN, INC., *Tokyo*
PRENTICE-HALL OF SOUTHEAST ASIA PTE. LTD., *Singapore*
EDITORA PRENTICE-HALL DO BRASIL, LTDA., *Rio de Janeiro*

To my wife—Carmen
My children—Dela, Tricia, and Lee
And the Grandbrats—Jorge and Mando

Contents

Contents

Preface

The use of computer-aided engineering analysis is mandatory in most engineering disciplines today. In fact, accrediting agencies are now requiring the integration of courses in computer analysis for engineering disciplines. Whether a student of engineering, a practicing engineer, or a technician, you will undoubtedly be required to use a program such as PSpice to perform such analyses.

Another aspect of using this type of program that has not been brought out in any of the other books that I have read is that it is fun! When you use a program such as PSpice, you are free to let your mind wander, to try anything that sounds reasonable—or for that matter, unreasonable. If you are curious as to what will happen "if I do this," go ahead and do it. You cannot burn out any expensive equipment, unless you have some very special and sensitive form of computer that gets insulted easily. Flights of fancy are perfectly acceptable when simulating.

It is assumed that the person using this book is a student who is presently taking at least 200-level circuits courses and has a math background that includes at least trigonometry and complex numbers. More advanced math is only a benefit for some of the problems in the last four chapters of the book—and then, not for all of them.

It is also assumed that the student has a passing acquaintance with the use of a computer and with the use of one of the many ASCII editors available. It is not reasonable to assume that everyone is using the same editor, and to present a chapter on the use of an editor.

For this book to be useful, it must provide the student, primarily, with a tutorial approach to circuit simulation. It should also provide an in-depth discussion of features such as PROBE and the models and subcircuits to be found in PSpice. This the book does. Further, the book should provide the student with ongoing support as he or she progresses through his or her education. This is also done in later chapters of the book. Any differences between student versions and more advanced versions of the program are pointed out wherever possible.

It is neither necessary nor necessarily desirable that this book be covered in one quarter or one semester. The book is meant to be an aide as the student progresses through the educational process and requires more advanced analyses. I hope that as students progress through their education and encounter more and more technical courses, they will find some portion of the book that covers a present area of interest.

Finally, it is time to thank those who helped make this book possible. First is my family. Your support and patience over this long haul have meant everything. Also my students, too many to mention by name. Finally, my co-teachers. If they didn't keep asking "can PSpice do this?" I might have missed something important.

L. H. Fenical

Introduction

WHAT IS SPICE?

SPICE is an acronym for Simulation Program with Integrated Circuit Emphasis. SPICE was developed at the University of California at Berkeley in the early 1970s. The original SPICE was evolved from a simulation program by the name of CANCER, also developed at UC–Berkeley. The basis of CANCER was a simulation program named ECAP, developed by IBM in the early 1960s. SPICE is a family of programs that are a culmination of efforts over about a 30-year span. It is a fact that SPICE is the most used simulation program in the world today. SPICE was developed using public money; thus the public is entitled to use it.

The most used version of SPICE today is SPICE2G.6, although there is another, SPICE3A.7. SPICE3A.7 is specifically designed to work within the computer-aided design (CAD) research program at UC–Berkeley. SPICE2G.6 is written in FORTRAN, and SPICE3A.7 is written in C.

WHAT IS SPICE USED FOR?

SPICE is a program whose function is to simulate the operation of a circuit without having to build the circuit with hardware. As such, you can use SPICE to evaluate component variations, temperature effects, noise levels, distortion, and other important circuit parameters. You can optimize circuit performance and predict, to some degree, circuit yield in manufacture. SPICE is capable of various analyses in the dc, frequency, and time domains. The operating point of devices can be simulated, the frequency response of passive and active circuits can be simulated, and the time response of passive and active circuits can be simulated.

SPICE contains models for resistors, capacitors, inductors, BJTs, JFETs, and other active devices. In all cases, if there are no parameters other than the most basic, SPICE assumes default values for all the parameters not specified. In all cases, the default parameters of SPICE provide a good working model for the devices being simulated.

WHAT DOES SPICE RUN ON?

Until recently, SPICE required a mainframe computer such as a VAX platform to run on. As a result, smaller companies that do not have a mainframe, or that rent one, cannot make use of SPICE. Most companies, including small ones, have PCs. Naturally since SPICE was developed using public money, this would lead to a spate of SPICE clones that run on a PC. There are several versions of SPICE that run on the PC, among them:

> *IS_SPICE,* from Intusoft in San Pedro, California. This is a version of SPICE2, and the conventions for SPICE must be observed. SPICE 2 requires that its input be in uppercase letters only, and .PRINT statements must come before .PLOT statements.
>
> *Z/SPICE* from ZTEC in College Place, Washington. This is a direct conversion of SPICE2.
>
> *ALLSPICE* from Contour Design Systems, Inc., Menlo Park, California (formerly ACOTECH). This SPICE version is no longer available.
>
> *PSpice* from MicroSim, Irvine, California. PSpice is a version of SPICE2 with many internal structure changes to improve convergence and simulation accuracy.

PSpice

PSpice by MicroSim, designed to work in the PC environment, was the first to use the IBM PC as a platform. Today, PSpice can be run on many platforms: the Sun workstation, VAX platforms, and Macintosh, to name a few. PSpice uses the same algorithms as SPICE and also conforms to the SPICE input syntax. PSpice is not case sensitive; the syntax can be either upper or lower case. PSpice has a graphics postprocessor called PROBE that can be used as a software oscilloscope. Waveforms can be viewed using simple commands, and many different "subanalyses," such as the input and output impedance of ac circuits and Fourier analysis can be done using PROBE. Neither SPICE nor PSpice is interactive. That is, if you wish to make a change in a circuit and observe its effect, you must edit the netlist of the circuit and rerun the program.

Several versions of PSpice are available.

Student version: the smallest version, available from Prentice Hall, Inc. It is limited by the fact that it comes on two 360K disks. It is still very powerful. Limitations are in the form of the types of functions that can be run, the number of semiconductors in a circuit is limited to 10, and the number of nodes that can be used is limited to 25. This version does not support the analog behavioral modeling option or the digital option. The student version of PSpice will run without a math coprocessor (5 to 15 times slower than with a coprocessor), but it is very slow for some of the analyses in this book. Also, the student version of PSpice will not run the Laplace and table functions of the analog behavioral modeling part of PSpice.

Evaluation version: limited in its nodes and semiconductor count to the same numbers as the student version, but more powerful than the student version in that it will run the analog behavioral modeling option of PSpice. The evaluation version also runs the digital option, in the mixed analog and digital mode or in the straight digital mode. The math coprocessor is mandatory in this version.

Full version: the most expensive but most powerful version. This version can handle up to 200 bipolar transistors, or 150 MOSFETs. The math coprocessor is mandatory for this version also.

The programs in this book were run on a 80286-based IBM-compatible computer running at 20 MHz, with a math coprocessor. In most cases a hard disk is not required, but there are some analyses that create a large amount of data. A hard disk is mandatory if these programs are to be run. Transient analysis programs are those most likely to produce large amounts of data and have long execution times. Copies of the PSpice student version can be obtained from Prentice Hall, Inc., Route 59 at Brook Hill Drive, West Nyack, NY 10995.

ORGANIZATION OF THE BOOK

It is assumed that most people using this book will be students. Thus the book starts with simple dc analysis, then progresses to ac analysis, then to passive and active device models, and finally, to subcircuits and transient analysis and integrated circuits. The later chapters of the book present more advanced subjects and the methods for analyzing circuits such as feedback control circuits and communication circuits.

Since nobody writes a book "the way I want it," this is simply mine. The only order of presentation that should be adhered to is in the presentation of dc and ac analysis in the first two chapters. The other areas can be taught in any order deemed necessary by the instructor. The first six chapters compose most of the tutorial in this book and should be taught in one quarter or one semester. The last six chapters will be useful to the student as they progress from one technical area to another. It is hoped that the student will be able to read and understand what is needed for the last six chapters.

A shortened version of the commands and devices is given in Appendix B. There are devices and commands in the appendix that are not covered in the book. However, after a quarter, or so, of instruction the student should be able to read about and use the devices not covered.

PSpice®: A Tutorial

Getting Started in PSpice

OBJECTIVES

1. To learn how to set up a circuit diagram for use in PSpice by adding node designations to the diagram.
2. To learn how the three basic passive components—the resistor, capacitor, and inductor—are applied in a PSpice netlist.
3. To learn how the two basic independent sources are applied in a PSpice netlist and the convention for currents in PSpice.
4. To learn how to make simple dc analyses of bias points for the circuits analyzed.

INTRODUCTION

There are several versions of PSpice. There is a student version; an evaluation version, which is more advanced than the student version; and the full version, which has all the functions of PSpice and a larger capacity for nodes and devices than either of the other two. If you are using either the student version or the evaluation version, you are limited to 10 semiconductor devices and 25 nodes in the circuit. In most cases this will be enough. When we need more devices than can be run in either the student or evaluation versions, such as when you need more than one operational amplifier (op amp), we will find ways around this obstacle. In all cases, the netlists in this book have all been run on the student version. Most have also been run on other versions as well. In every case, the outputs have been the same. When we need them, the differences between the versions will be annotated.

The purpose of PSpice is not to design circuits, but rather, to analyze the operation of a circuit that you have designed. This allows you to ''construct'' a circuit using PSpice's software and to test the circuit before it is constructed using hardware. In doing this, PSpice allows us to test the limitations of a circuit before it is built, giving the designer the option of changing the circuit before the physical prototype stage. To accomplish this, you must design the circuit beforehand and place the parameters in a PSpice netlist in a form usable by the program. Many different parameters may be tested for, but for the moment, only the dc (direct current) parameters of circuits will be investigated. As the need for more types of parameters, such as frequency and pulses, is required, they will be introduced. The purpose of this chapter is to present you with the principles used to set up circuit files (netlists) for PSpice. You will also learn to label the nodes in a circuit, what the required syntax for devices and sources is, and then to apply the commands and statements of PSpice to analyze the circuit.

1.1 TEXT FILES FOR PSPICE

To use PSpice, it is necessary first to create a file for the circuit that you have designed. This file is called a *netlist*. PSpice is the program. The input file (netlist) is not capable of running independently; thus it is not a program. The editor used must be capable of producing an ASCII file. If you are using the student version, an editor such as EDLIN, or one of the many editors available either from your own sources or from a bulletin board, will work. If you are using the advanced versions of PSpice, there is a text editor in the interactive shell. The only restriction is that the editor must produce an ASCII file.

All of the netlists in this book have been run using the student version of PSpice. The editor used in this book is one from the public domain.

1.2. REQUIREMENTS FOR CREATING THE TEXT FILE

In PSpice, as in any other analysis program, it is necessary to use standardized names for the devices and sources used by the program to calculate the output that we want. These component names must be used in the text file that is created or the program will not be able to calculate the circuit parameters. In this chapter we will be using only a few devices and two independent source types. These are all that we need to remember at this time. The basic devices and sources are listed in Table 1-1. There are more devices available, and they will be introduced as they are needed. For the moment we will use only the resistor and the two types of source shown.

When we write a netlist, we must know and use the proper syntax for the devices and sources that we will be using. In the simple resistive circuits of this chapter, only

TABLE 1-1

Devices		Independent sources	
Resistor	− R⟨name⟩	Voltage source	V⟨name⟩
Inductor	− L⟨name⟩	Current source	I⟨name⟩
Capacitor	− C⟨name⟩		
Diode	− D⟨name⟩		
Transistor	− Q⟨name⟩		
FET	− J⟨name⟩		

voltage and current sources and resistances are used. The general syntax for the voltage or current source and resistor is

```
V<name> < + node> < - node> DC <voltage value>
I<name> < + node> < - node> DC <current value>
R<name> < + node> < - node> <resistance value>
```

The ⟨name⟩ shown after the abbreviations are names that uniquely identify the device or source. The devices and voltage and current sources may be given names that specify the function they perform. VIN is an example of naming a source that is an input or supply voltage to the circuit that you have designed. Sources and devices are separate entities; that is, a device is a passive or active component of the circuit that is used for biasing, amplification, or other purposes, while the sources supply power to the devices. For this reason they are called *independent sources*. For dc analysis in general, the sources provide the power for operating the circuit and the input signals to some of the controlled circuits. We shall see the application of the sources as both biasing and input voltages in this and the next chapters.

PSpice recognizes only three general structures in its format:

1. *Command statements.* These statements are always preceded by a dot, or period.
2. *Sources.* For the moment, only independent sources of voltage and current are considered.
3. *Devices.* For the moment, only passive devices are considered.

An example of the command statement format is

```
.END
```

which is how we end all netlists. All command statements must be preceded by a dot or period. An example of a device statement is

```
R121  1  2  1000OHMS
```

This is a resistor connected between two nodes, 1 and 2, with a value of 1000 ohms (Ω). An independent source would be

VIN 1 0 DC 12VOLTS

or

Ibias 0 5 DC 1(AMP)

For now, these are the only types of devices and statements that will be contained in the PSpice netlists that we create. To create a netlist, we must first design the circuit and then assign nodes to it so that the program can analyze the circuit. Nodes are connection points in the circuit. We discuss this in some detail later. But first, it is necessary to understand how PSpice handles the values of the components and sources that are used.

1.3. COMPONENT VALUES IN PSPICE

Component values can be stated several different ways in PSpice. You know that a number such as 0.0039 can be expressed as

0.0039   3.9E − 3   3900E − 6   3.9m

PSpice allows this, too, but in addition has nine standard suffix letters that may be used for numeric quantities. These are shown in Table 1-2. Notice that all of the suffixes may be written in both uppercase and lowercase letters. PSpice is not sensitive to upper or lower case. SPICE is—it allows only uppercase letters. It is for this reason that if the megohm is the required quantity, we must type the three letters meg or MEG. Lowercase m is traditionally used to denote milli, and uppercase M is used to mean mega. In PSpice both upper- and lowercase M (m) are used for milli—thus the need to

TABLE 1-2

Suffix	Prefix represented	Value
F, f	femto	10^{-15}
P, p	pico	10^{-12}
N, n	nano	10^{-9}
U, u	micro	10^{-6}
M, m	milli	10^{-3}
K, k	kilo	10^{3}
MEG, meg	mega	10^{6}
G, g	giga	10^{9}
T, t	tera	10^{12}

distinguish between the two. The letter suffixes shown in Table 1-2 can be used to save a few keystrokes.

A note of caution regarding use of the suffixes is in order here. You may have noticed that the names *ohms, volts,* and *amps* (amperes) were used with the values. This is not necessary to do, since PSpice knows that resistors are in ohms, voltages are in volts, and so on. If you decide to use the names with the values, no space may be shown between the value and its name. PSpice is a field-oriented program, in which spaces or commas represent the end of a field. Thus a space between the value and its units tells PSpice that a new field is following the space. This will cause no end of problems since nodes do not have to be numbers. You may also have noticed the parentheses around the word "AMP." In the student version that I have, A or a means atto, or 10^{-18}. Further, if you were to define a 10-microfarad (μF) capacitor as

$$1E-5FARAD$$

you may be surprised to find that you now have a capacitance of $1E - 21$ farads. If a unit name is one of the standard suffixes, it is best not to try to use it in this application.

1.4. CIRCUIT NODES

A *node* is any point in the circuit where two or more wires would come together. That is, if you were building the circuit, anywhere that you would connect two or more components together is a node. All nodes, regardless of what is connected to them, must have a dc path to ground. The order of the nodes is not generally important, but some semblance of order should be used to make the circuit file easy to trace to the circuit being analyzed.

The nodes of a circuit are labeled as shown in Figure 1-1. There are only three nodes in this circuit. Two nodes connect the resistors and the voltage source together,

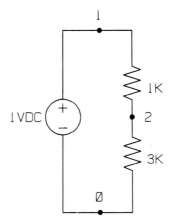

Figure 1-1

and the third node is the ground node. The ground node is always labeled as the zero node. It is important to emphasize that the zero node is reserved for the ground, or reference, node only! The zero node is also global in its application. When we get into more complex circuits, we will see that the zero node applies to all circuits as well as those called subcircuits that are stored in a library file elsewhere on the disk. All nodes must have a dc path to the zero node.

For the simpler circuits that we shall do it is wise to follow this procedure: Calculations for the circuit should be made before the netlist is written, and then the results of your calculations and the PSpice calculations should be compared after the netlist is run. Since this is your first exposure to running PSpice, it is a good idea to know beforehand the correct values for the bias, the current, and the power dissipation of the circuit.

1.5. CREATING THE NETLIST

To make a netlist for any analysis, we must first know how the components are connected in the circuit and how the nodes are labeled. For this chapter we are using only two terminal devices, so there are only two connections that need to be defined for each device.

There are some simple rules that must be followed when making a netlist for any analysis, and these rules are:

1. All of the devices in the circuit are considered to be connected together by nodes, just as you would solder their leads together.
2. The nodes must be either positive numbers, usually integers, or names. It is possible that the nodes may be named by numbers with a decimal, but for most netlists this is probably neither necessary nor desirable. Nodes may also have meaningful names within the circuit, such as IN, OUT, AMP1, and so on. It is recommended that node names be kept to no more than eight letters or numbers, even though more may be used (a name may be up to 131 characters long).
3. Zero is reserved by PSpice for the ground (reference) node.
4. All circuits must have a zero (0) node as a reference, and every other node in the circuit must have a dc current path to this reference. (This is an absolute requirement and may not be ignored.)
5. PSpice requires that all nodes be connected to at least one other node; this avoids open circuits and is also an absolute that may not be ignored. If an output must simulate an open circuit, a very large resistance [1 gigaohm (GΩ) or more] can be connected from the output to ground.
6. The simplest syntax for the two-terminal device is

    (Device[R,C,L,V,I]<NAME>) <+NODE> <-NODE> <VALUE>

The device part of the command defines what component is being used: a resistor, capacitor, inductor, source, and so on. After the letter indicating what the device is may come other letters or numbers to identify the device or source uniquely. It is not required that a specific method, other than the first letter, be used to name the components. But the same name may not be used for two components in the circuit, even if they are the same type of component. Nodes may also have names rather than numbers. It is only important that the names or numbers uniquely identify the node or component. Names usually should be kept to a maximum of about eight characters.

7. All voltage and current sources are treated as perfect. The perfect voltage source has zero internal impedance. This allows you, as the designer, to use the actual source resistance around which you are designing. It is always wise to insert some source resistance in the circuit for the voltage source even if none is required. The perfect current source has infinite internal impedance and therefore does not qualify as a path to ground. It is necessary that some parallel resistance be added to the current source since any node that does not have its own ground return to which the current source is connected will be floating. This resistance may be as large or small as necessary so as not to disturb the operation of the circuit.

8. All netlists must start with a title line and end with the statement .END. The title line may be a blank line if you wish, but it is usually more convenient to use a title. This is mandatory and may not be ignored. If no title line is present, the first netlist line becomes the title line.

9. As in all software, the file must have a name and an extension. The files we create will have the extension, .CIR, but may have other extensions. We discuss this further later.

1.6. THE CIRCUITS

The circuits used for this chapter are simple dc circuits, examples of which can be found in any text on electric circuits. We will progress from simple circuits, such as the voltage divider, to more complex circuits using more than one independent voltage or current source. You should also work up some circuits that you can analyze to get practice on writing netlists for PSpice. All of the netlists in this book, except a few in Chapter 12, have been run on the student version of PSpice. The output from all the netlists that are run using the netlists in this book should be identical to the ones shown in this book.

Example 1. Resistive voltage divider

The first of the circuits to be designed is the resistive voltage divider. Although it may seem trivial, this circuit was, and still is, used to provide low current voltage supplies for circuits in some radios and TV sets. The circuit will be broken down into the equations needed, then the design will be drawn for PSpice and input to a file.

The simplest resistive voltage divider is made up of two resistors and a voltage source. The resistors may be labeled R_1 and R_2. The quantities needed to calculate the resistive voltage divider are the input voltage and the resistor values. Also, if we know any three of the four values of the circuit, the remaining one may be found. The equations for this circuit are as follows:

$$V_{\text{out}} = V_{\text{in}} \frac{R_2}{R_1 + R_2} \tag{1}$$

$$R_2 = \frac{V_{\text{out}} R_1}{V_{\text{in}} - V_{\text{out}}} \tag{2}$$

$$R_1 = R_2 \frac{V_{\text{in}} - V_{\text{out}}}{V_{\text{out}}} \tag{3}$$

$$P_T = I^2 \times R_T \tag{4}$$

For this circuit refer to Figure 1-2. The voltage across R_2 is readily seen to be 0.75 volt (V), the current for the circuit is 0.25 milliampere (mA), and the power dissipated by the circuit is 0.25 milliwatt (mW).

Before we can do anything with this circuit, we must put it in a form acceptable to PSpice. To do this we must first determine the nodes in the circuit. The netlist for the resistive divider is shown below. The only absolute requirement for the netlist is that the netlist start with a title line and end with the .END command line. The lines for devices, sources, and command lines may be in any order between these two lines. We could put the voltage source at the bottom of the list, or between the two resistors, it does not matter. Just remember to include any periods before commands.

When you have typed this circuit into your editor, save it under the name RDIV.CIR. The circuit file should be saved on the same disk that contains the PSpice program startup file. At this time it is important to point out that the name of a netlist file may contain up to eight characters, both letters and numbers, but should begin with a letter.

```
*RESISTIVE DIVIDER CIRCUIT
VIN 1 0 1.0VOLT
R1 1 2 1.0KOHM
R2 2 0 3.0KOHM
.END
```

What PSpice will do with this circuit is calculate the bias point for the resistors, the voltage at all nodes, the current in the circuit, and the power dissipation in the circuit. They are then listed in an output file with the same name as the circuit file but with the extension .OUT (this is a default). The output file is not required to have the same name as the circuit file, but this is usually convenient. The method of giving the output file another name will be covered later.

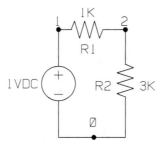

Figure 1-2

When the current is calculated for this circuit, it will be shown to be negative (current flows from the positive terminal, through the source, to the negative terminal of the source), so the current is printed as a negative value. For most current sources, it is convenient to list them with node zero first, then going to another node to make the current produce positive voltage drops. Let us analyze this netlist line by line so that we have an idea of what is going on. Although simple, this netlist uses all the rules of Section 1.5. The same principles apply to all the netlists we write.

Line 1:

*RESISTIVE DIVIDER CIRCUIT ANALYSIS

This line is the title line for the netlist. All netlists must start with a title line. This line will show on the screen when the netlist is executing the bias point or other analysis. If this line is inadvertently omitted, the first line of the netlist becomes the title line, and an error will be reported. Line numbers are usually not required by an editor (an exception is EDLIN, the DOS editor). The use of all capital letters in the netlist is also not mandatory in PSpice, but in SPICE, capitals are mandatory. The use of capitals in PSpice is just a good habit, and may be convenient for reading on the screen of the CRT. The use of an asterisk in the first column of a line makes the line a comment line that will be ignored by PSpice. The use of the asterisk in the title line is not necessary but will cause no problems for the program. If a comment is needed (or desired) after a device line, the semicolon is used. This indicates that the information following the semicolon is a comment. You will see this and the asterisk used as needed in the netlists in this book. Generous commentary within a netlist is always good for memory jogging six months after you have written the netlist and no longer remember what went through your mind when you wrote the netlist.

Line 2:

VIN 1 0 1.0VOLT

This line specifies the voltage source for the system. In this line, all of the possible default values for a voltage source have been used. The total voltage specification includes the fact that the source is a dc source; the default is dc so we do not have to specify it here. Since we are only interested in the dc parameters for this analysis, we allow all of the other parameters to default. We may also note here that if the voltage source is reversed, that is, if it is listed as

VIN 0 1 1.0VOLT

the voltage will be negative rather than positive. This is due to the current convention previously noted. We could also have made the voltage negative by using the notation first shown and showing the voltage as − 1.0VOLT.

VIN is the name we will use for the voltage source. The voltage source is connected between common, node 0, and the node labeled 1. The value of the voltage at the source is 1.0 V. The decimal value is not required, but may be used. The word VOLT is also not a required quantity. PSpice understands that the value is in volts by the fact that a voltage source, V, was specified. If the word VOLT is used, there must be no space between the value and the name of the quantity, since PSpice uses spaces to separate the field values. A space between the value and the name of the quantity would be interpreted as an additional field by PSpice and an error would result.

Line 3:

$$R1 \quad 1 \quad 2 \quad 1.0KOHM$$

This line connects the voltage source at node 1 to node 2 through R_1. The value of R_1's resistance is 1000 Ω. Note that we are using K instead of either powers of 10 or the actual value for the resistance.

Node 1 now has the two connections required but we still have not completed the current path to the reference (ground) node. This will be done in line 4.

Line 4:

$$R2 \quad 2 \quad 0 \quad 3.0KOHM$$

This line completes the circuit and makes the current path to ground complete. The requirements for this line are the same as those for line 3. Node 2 now has two connections, as does node 0, which fulfills the requirement that there be no open circuits anywhere in the circuit. Also, we do not have to supply the voltage source with any resistance for this netlist. If we wish, we could consider R_1 to be the source resistance for the circuit.

Line 5:

$$.END$$

This is how PSpice knows that any file has ended. It is mandatory! If the .END statement is not used, the program will report an error and abort. The error will probably be "END CARD EXPECTED." The device and command lines may be in any order between the starting title line and this line. The title line and the .END statement must be the first and last lines in the netlist.

1.7. RUNNING PSpice

In general, to run PSpice and the netlist just created, it is necessary only to type PSpice and the name of the netlist we have created. Path commands for the disk drives are also allowed in the startup command. If you have entered the netlist and saved it, run the netlist by typing

$$PSPICE \quad RDIV$$

if you are using the student version, or

$$PS \quad RDIV$$

if you are using one of the advanced versions that run in a shell. Note that the extension .CIR is not necessary for the program to run and select the proper file. If you are using a hard disk and have the PSpice batch file in one drive and the circuit files in a different section of the disk under a different drive name, type

```
PSPICE    <drive>:RDIV
```

or

```
PS    <drive>:RDIV
```

If you should want to give the circuit file a different extension, this is allowed, but the extension must be included in the filename when the netlist is run.

On startup you can include the circuit file name and designate the output file to which you wish the data to be written. It is not mandatory that the circuit input and output file names be the same, nor their extensions. You may include a drive specification for the output file, if desired. The general syntax for the startup procedure is

```
PSPICE [PS] <DRIVE SPEC.>:<INPUT FILENAME.EXT> <DRIVE SPEC>:
+                                    <OUTPUT FILENAME.EXT>
```

If you do not specify an input filename when calling the program, the program will stop and ask for both an input and an output filename. It is only mandatory that the input filename be given. If you want the output file to have the same name as the input file, simply press return and this will happen. If no output file is named, a file is created that has the same name as the circuit file, with the extension .OUT. Thus as our circuit runs under the name RDIV.CIR, a file named RDIV.OUT is created. The output file contains the results of bias calculations made on the circuit. For the student version of PSpice, the output file must be manually printed to the screen. For the advanced versions use the browse option in the file menu. The general form for accessing an output file is

```
TYPE <FILENAME.OUT>
```

For this circuit the output data are accessed by typing

```
TYPE RDIV.OUT
```

An example of the data for this circuit is shown in Analysis 1-1. For large netlists a large amount of data is generated. Once you have run the netlist and used the data, it is wise to use a separate disk to store the netlist and output files for future reference. In general, output data are generated any time that a circuit netlist is run, so it is not necessary to keep all of the output files you have generated.

Example 2. Resistive bridge circuit

The circuit to be analyzed by PSpice is shown in Figure 1-3. The analysis for this type of circuit is much more complex than that for the simple resistive divider. We can use mesh analysis and Kirchhoff's laws to analyze the bridge. Notice that there are four nodes that

```
******* 01/04/80 ******* Evaluation PSpice (January 1990) ******* 16:11:05 *******

*RESISTIVE DIVIDER CIRCUIT - ANALYSIS 1-1

****      CIRCUIT DESCRIPTION

*********************************************************************


VIN 1 0 1.0VOLT
R1 1 2 1.0KOHM
R2 2 0 3.0KOHM
.options nopage
.END


****      SMALL SIGNAL BIAS SOLUTION        TEMPERATURE =    27.000 DEG C


NODE    VOLTAGE      NODE    VOLTAGE      NODE    VOLTAGE      NODE    VOLTAGE

(    1)    1.0000  (     2)     .7500



    VOLTAGE SOURCE CURRENTS
    NAME          CURRENT

    VIN         -2.500E-04

    TOTAL POWER DISSIPATION   2.50E-04  WATTS


      JOB CONCLUDED

      TOTAL JOB TIME            6.04
```
                                    Analysis 1-1

will have voltages in reference to ground, and a single pair of nodes that may, or may not, have a voltage difference between them. For this analysis we consider the following resistor values and input voltage: $R_1 = 10\ \Omega$, $R_2 = 25\ \Omega$, $R_3 = 10\ \Omega$, $R_4 = 20\ \Omega$, $R_5 = 50\ \Omega$, $R_6 = 100\ \Omega$, and $V_{in} = 10$ V.

   It is obvious that there are three mesh currents to be solved for in the bridge circuit, and these currents—i_1, i_2, and i_3—are identified as I1, I2, and I3. The mesh equations for these currents are

$$i_1(R_1 + R_2 + R_3) - i_2R_2 - i_3R_3 = 10 \tag{5}$$

$$-i_1R_2 + i_2(R_2 + R_4 + R_6) - i_3R_6 = 0 \tag{6}$$

$$-i_1R_3 - i_2R_6 + i_3(R_3 + R_5 + R_6) = 0 \tag{7}$$

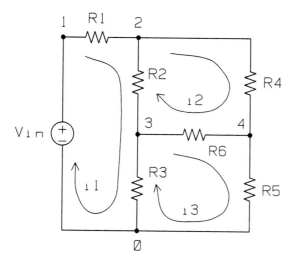

Figure 1-3

The node voltages calculated by PSpice are listed in the output file. Once again, if you are using the student version, you will need to use TYPE ⟨filename⟩ .OUT to bring the output file to the screen. The voltages at the nodes are solved for by adding up the currents and then multiplying by the resistance from the node to ground, or node to node. You should calculate the values for this circuit and then type in the netlist, run it, and compare the results that you have calculated to the results calculated by PSpice. It is important that you understand that the node voltages are with respect to ground, not from node to node. We will learn to make node to node measurements in Chapter 2. Thus your calculations should be for the node voltage to ground also. There should be no difference between your calculated values and the values that are calculated by PSpice. The netlist for the bridge circuit is

```
*RESISTIVE BRIDGE DEMO CIRCUIT
VIN 1 0 10Volts
R1 1 2 10
R2 2 3 25
R3 3 0 10
R4 2 4 20
R5 4 0 50
R6 3 4 100
.END
```

The output for the bridge circuit is shown in Analysis 1-2.

******* 01/04/80 ******* Evaluation PSpice (January 1990) ******* 16:11:51 *******

*RESISTIVE BRIDGE DEMO CIRCUIT - ANALYSIS 1-2

****   CIRCUIT DESCRIPTION

****

```
VIN 1 0 10VOLTS
R1 1 2 10OHMS
R2 2 3 25OHMS
R3 3 0 10OHMS
R4 2 4 20OHMS
R5 4 0 50OHMS
R6 3 4 100OHMS
.options nopage
.END
```

****   SMALL SIGNAL BIAS SOLUTION   TEMPERATURE =   27.000 DEG C

NODE	VOLTAGE	NODE	VOLTAGE	NODE	VOLTAGE	NODE	VOLTAGE
(1)	10.0000	(2)	6.9267	(3)	2.1537	(4)	4.5984

```
    VOLTAGE SOURCE CURRENTS
    NAME           CURRENT

    VIN          -3.073E-01

    TOTAL POWER DISSIPATION   3.07E+00  WATTS


        JOB CONCLUDED

        TOTAL JOB TIME          6.21
```

Analysis 1-2

1.8. SOME ADDITIONAL CIRCUITS FOR ANALYSIS

Here are some additional circuits and netlists to be analyzed using various techniques, such as superposition. The circuits use various combinations of voltage and current sources and all have the nodes identified.

Example 3. Two-voltage-source circuit

The circuit for this example is shown in Figure 1-4. The netlist for this circuit is

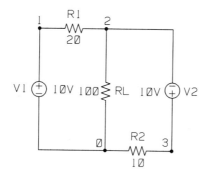

Figure 1-4

```
*TWO VOLTAGE SOURCE PROGRAM.
V1  1  0  10
R1  1  2  20
RL  2  0  100
V2  3  2  10
R2  3  0  10
.END
```

The output is shown in Analysis 1-3.

******* 01/04/80 ******* Evaluation PSpice (January 1990) ******* 16:12:37 *******

*TWO VOLTAGE SOURCE PROGRAM - ANALYSIS 1-3

****   CIRCUIT DESCRIPTION

****

```
V1 1 0 10
R1 1 2 20
RL 2 0 100
V2 2 3 10
R2 3 0 10
.options nopage
.END
```

****   SMALL SIGNAL BIAS SOLUTION   TEMPERATURE =   27.000 DEG C

NODE	VOLTAGE	NODE	VOLTAGE	NODE	VOLTAGE	NODE	VOLTAGE
(1)	10.0000	(2)	9.3750	(3)	-.6250		

Analysis 1-3

```
VOLTAGE SOURCE CURRENTS
NAME          CURRENT

V1          -3.125E-02
V2          -6.250E-02

TOTAL POWER DISSIPATION   9.38E-01  WATTS


    JOB CONCLUDED

    TOTAL JOB TIME          4.73
```
Analysis 1-3 (cont'd)

Example 4. Two-current-source circuit

The circuit for this example is shown in Figure 1-5. The netlist for this circuit is

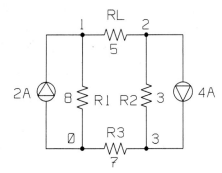

Figure 1-5

```
*TWO CURRENT SOURCE CIRCUIT
I1  0  1  2AMPS
R1  1  0  8
RL  1  2  5
R2  2  3  3
I2  2  3  4AMPS
R3  3  0  7
.END
```

The output is shown in Analysis 1-4.

```
****** 01/04/80 ****** Evaluation PSpice (January 1990) ****** 16:13:22 ******

*TWO CURRENT SOURCE CIRCUIT - ANALYSIS 1-4

****     CIRCUIT DESCRIPTION


********************************************************************************

I1 0 1 2AMPS
R1 1 0 8
RL 1 2 5
R2 2 3 3
I2 3 2 4AMPS
R3 3 0 7
.options nopage
.end


****     SMALL SIGNAL BIAS SOLUTION       TEMPERATURE =   27.000 DEG C


NODE   VOLTAGE     NODE   VOLTAGE     NODE   VOLTAGE     NODE   VOLTAGE

(   1)   14.6090  (   2)   13.7390  (   3)    1.2174



   VOLTAGE SOURCE CURRENTS
   NAME            CURRENT

   TOTAL POWER DISSIPATION   0.00E+00  WATTS


      JOB CONCLUDED

      TOTAL JOB TIME             6.04
```

Analysis 1-4

Example 5. Voltage and current source circuit

The circuit for this example is shown in Figure 1-6. The netlist for this circuit is

```
*VOLTAGE AND CURRENT SOURCE CIRCUIT.
V1 1 0 10
R1 1 2 25
R2 2 3 100
I1 2 3 .08AMP
RL 3 0 50
R3 3 4 40
V2 0 4 12
.END
```

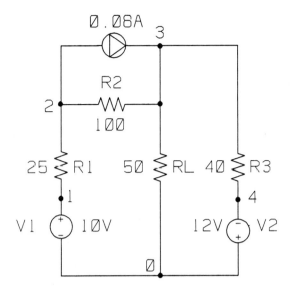

Figure 1-6

The output is shown in Analysis 1-5.

```
******* 01/04/80 ******* Evaluation PSpice (January 1990) ******* 16:14:10 *******

*VOLTAGE AND CURRENT SOURCE CIRCUIT ANALYSIS 1-5

****      CIRCUIT DESCRIPTION

**********************************************************************************

V1 1 0 10
R1 1 2 25
R2 2 3 100
I1 3 2 0.08AMP
RL 3 0 50
R3 3 4 40
V2 4 0 12
.options nopage
.END

****      SMALL SIGNAL BIAS SOLUTION      TEMPERATURE =   27.000 DEG C

NODE   VOLTAGE      NODE   VOLTAGE      NODE   VOLTAGE      NODE   VOLTAGE

(    1)   10.0000  (    2)   10.7920  (    3)    5.9623  (    4)   12.0000
```

Analysis 1-5

```
VOLTAGE SOURCE CURRENTS
NAME           CURRENT

V1             3.170E-02
V2            -1.509E-01

TOTAL POWER DISSIPATION    1.49E+00   WATTS


    JOB CONCLUDED

    TOTAL JOB TIME          6.04
```
                                   Analysis 1-5 (cont'd)

1.9. BATCH-FILE PROCESSING USING PSPICE

A useful function of PSpice is its ability to run netlist files as a group, called a *batch*. When PSpice has run an analysis, it looks for a title line after the .END statement for the netlist it has just run. If the program finds neither a blank nor a title line, the program terminates. If a blank, or title line is found, PSpice looks for command and device statements and will execute the analysis specified. The program then does the same procedure over again. PSpice will do this as many times as there are files to be run. For this reason, if you inadvertently leave a blank line below the .END statement, the output file will contain a message indicating that there are no devices in the file to be analyzed.

SUMMARY

There are a few types of PSpice available: the student version, the evaluation version, and the full version. The evaluation version and the full version operate in a shell environment.

A circuit to be analyzed must have nodes assigned to the connection points of the circuit components. In this chapter we have only dealt with three fundamental components: the resistor, the capacitor, and the inductor. Further, we have dealt with the two independent sources available in PSpice, the voltage source and the current source. All the nodes of the circuit must have at least two connections and a path to the reference node (ground).

For the program to analyze a circuit, PSpice requires a netlist. The netlist can be written using any ASCII editor. The netlist must always start with a title line and end with the .END command line. The components and sources in the netlist may be in any order between these two lines.

PSpice has several standard suffixes that can be used in place of the numeric value of the component.

The results of a dc bias point analysis are written to a file called the dc output file.

PSpice can run a group of files called a batch.

SELF-EVALUATION

For the circuits shown in Figures P1-1 to P1-5, the zero node is identified. Assign nodes to the circuit and find the voltages at each node. Verify the values provided by PSpice by calculating the voltages yourself. This is an opportunity to try PSpice as a batch-file processor. To do this all that is required is to put the netlist files one after another after each .END command.

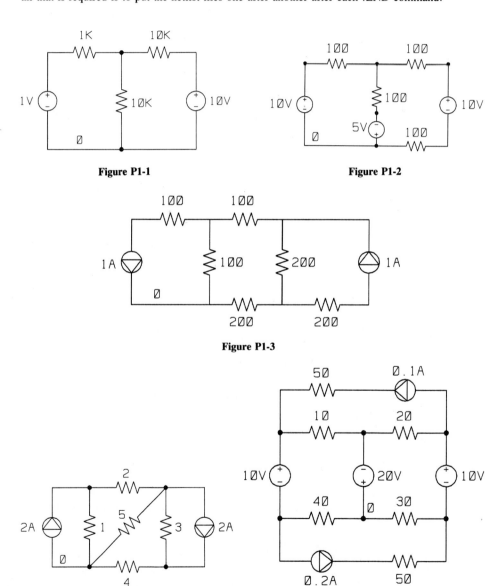

Figure P1-1

Figure P1-2

Figure P1-3

Figure P1-4

Figure P1-5

2

DC and AC Sweep; Z_{in} and Z_{out} of Passive Circuits

OBJECTIVES

1. To learn the methods by which the four types of dc sweep can be used in a circuit file.
2. To learn the methods by which the three types of ac sweep can be used in a circuit file.
3. To learn to use the .PRINT and .PLOT commands to store data to the output file for a circuit and to plot text graphics in the output file.
4. To learn to use the graphics postprocessor PROBE for high-resolution graphics.
5. To learn how to use PSpice to calculate the dc and ac input and output impedance of passive circuits.

INTRODUCTION

In Chapter 1 we investigated how to label nodes and print to the screen the information calculated for a circuit. This information was contained in the output file created by PSpice. Although this information is valuable, most of the time we will not be working with simple circuits. More information is needed than bias point calculations supply. In this chapter we investigate some of the functions available in PSpice that allow us to get a range of values for various inputs. These functions come under the general heading of sweep functions. Two types of sweep are available: dc sweep and ac sweep.

2.1. .DC; SWEEPING DC VOLTAGES

This function allows us to change the value of a dc voltage source in a predetermined way. Along with this are three commands that we can use to store and display the information generated by the sweep:

1. .PRINT   produces tabulated data in the output file.
2. .PLOT   produces graphic output from the output file.
3. .PROBE   used for high-resolution graphics. Allows calculation of circuit parameters.

Each of these functions has certain specific requirements for its use, and some default values. All may be printed out as hard copy. Let's start with the dc sweep function declaration. The syntax for this function is shown below and must always be in this format or an error will be reported by the program.

```
.DC <sweep type> <source name>     <Start value> <Stop value>
+ <Increment value>                ;continuation of first line
+ <nested sweep function>          ;continuation of first line
```

Note the use of the semicolon in the two lines with plus signs. This is how you make a comment after a command or device line.

The command .DC indicates to the program that a dc voltage source will be moved through a range of values, starting with one value and ending at another value, with an increment that you specify. As an example, we could sweep the input voltage of the simple voltage divider through a range of say, 0 to 10 V in 0.1-V increments. Then calculate the output voltage across node 2 for each of these values. To do this it is not necessary to change any of the lines of the netlist. During the sweep portion of the program the sweep source value will override the fixed value in the netlist.

The .DC function gives flexibility in investigating a circuit, but it must be used with the commands listed above. If added to the circuit all by itself, the .DC function will run, but it will do the calculations and then throw them away. The output file will be empty! Because of this the .PRINT and .PLOT functions are used with the .DC command to retain and display the data that are generated. We discuss these functions in detail in the next section. First we discuss some methods of using the dc sweep that make this command so useful. These are, sweeping nonlinearly, sweeping a list of values and nesting sweeps.

There are four .DC sweep function types:

1. Linear
2. Decade
3. Octave
4. List

The four types are described below together with the general syntax for each of the types of sweep.

1. LIN. Linear sweep is the default sweep. It sweeps in the increments that are set by you. It is not required that you specify the command LIN in this type of sweep, as it is the default value. The general syntax for this sweep is

```
.DC LIN <sweep variable name> <start value> <end value>
+        <increment value>
+        <nested sweep>
```

2. DEC. Decade sweep is a logarithmic sweep that is done in powers of 10 (a 10:1 ratio). You must specify DEC for this sweep. Since this sweep is logarithmic, a value of zero is not allowed for its axis. The general syntax for the decade sweep is

```
.DC DEC <sweep variable name> <start value> <end value>
+        <points per decade>
+        <nested sweep>
```

3. OCT. Octave sweep is also logarithmic, but is done in powers of 2. This type of sweep must also be specified, and a value of zero is also not allowed. The general syntax for the octave sweep is

```
.DC OCT <sweep variable name> <start value> <end value>
+        <points per octave>
+        <nested sweep>
```

4. LIST. This function allows the calculation of specific points. Only the points that are listed are calculated. The general syntax for the list sweep is

```
.DC <sweep variable name> LIST <values in list>
+    <nested sweep>
```

Note that in this type of sweep, the sweep variable name comes before the keyword LIST. There is no actual sweep of the form taken in the other sweep types. Only the listed values are calculated. The sweep variable is set to each of the numbers in the list and calculated for that value.

Following are some examples of how the sweep lines may be written.

1. *Linear sweep:*

```
.DC LIN VIN 0V 10V 0.1V
.DC VIN 0V 10V 0.1V
```

This is a linear sweep of the input voltage starting at 0 V, ending at 10 V, with 0.1-V increments. The word LIN may be left out of the command syntax, as shown in the second line, since it is the default sweep.

2. *Decade sweep:*

```
.DC DEC Iin 1E-5 1E-3 20
```

The decade sweep is a power-of-10 sweep. The device being swept in this case is a current generator, I_{in}, and it is swept from 10 μA to 1 mA logarithmically. Last is the number of points that the sweep uses per decade of change, in this case 20. Since this is a logarithmic type of sweep, the value zero may not be used as one of the sweep values. You cannot, for instance, sweep between -10 μA and 10 μA since this value passes through zero. Only the number of points per decade is decided by you, not the calculation points.

    3. *Octave sweep:*

```
.DC OCT VIN 1 16 5
```

The same rules apply to the octave sweep as apply to the decade sweep. The octave sweep is another form of logarithmic sweep, and zero is forbidden in this type of sweep also. This sweep is done in powers of 2. You should be careful about the number of points swept per octave since there are many octaves if you are sweeping a large range of voltage.

    4. *List:*

```
.DC VIN LIST 0 5 -10 15 20 -25 -5 10 -15 -20 25
```

Positive and negative values can be mixed. Note that in this case the sweep variable is named before the word LIST.

    5. *Nesting sweeps:* Another function of the dc sweep that is very useful is the ability to nest sweeps, each with its own range. This function works quite similarly to loops in program languages; that is, it completes one inner loop for each increment of an outer loop. An example of a nested sweep is

```
.DC LIN V1 0V 10V 1.0V  V2 0V 5V 0.1V
```
                                                                        nested sweep

In this case, both sweeps are linear and the entire V1 sweep will be done each time the V2 sweep is incremented. Note that in PSpice, the second sweep is considered to be the outer loop, while the first sweep is the inner loop. V1 runs through its entire range before V2 is incremented.

2.2. .PRINT FOR STORING OUTPUT DATA GENERATED BY .DC

The sweep calculates values of data, and the data generated by the sweep command are stored in the output file if the .PRINT command is used in the netlist for the circuit. The columns of this table are in the order that you specify in the .PRINT command. All node voltages and currents can be printed, or only one. The source to be swept is always listed in the leftmost column of the output file. You must specify what is printed to the file, but you cannot do any calculations in the output list. You could not show power as a function of a node voltage multiplied by a device current.

The syntax of the .PRINT command is

```
.PRINT DC <output value> · · ·
```

There may be as many output values as are needed for the circuit.

As an example of the use of the sweep and print commands, let us add both commands to the bridge circuit netlist of Chapter 1.

```
*RESISTIVE BRIDGE DEMO CIRCUIT
VIN 1 0 10VOLTS
R1 1 2 10OHMS
R2 2 3 25OHMS
R3 3 0 10OHMS
R4 2 4 20OHMS
R5 4 0 50OHMS
R6 3 4 10OHMS
.DC VIN 0 10 .25 ;Linear sweep is the default used
.PRINT DC V(3) V(4) I(R6) V(3,4) ;data to be stored
.END
```

The .PRINT command causes the voltage at the nodes shown to be listed in the .OUT file. Notice the difference between the way the voltages and the current are called for. In PSpice, voltages are taken either referenced to ground, such as V(3) above, or between two nodes, such as V(3,4) above. Current, in PSpice, is always specified by the device through which it flows. The table set up by the .PRINT command is in the specific order that you list in the command. Thus this table will be in the following order:

```
Vin    V(3)    V(4)    I(R6)    V(3,4)
```

For this netlist we are generating 41 values for each parameter in the table. (Zero to nine is 10 values, with four values for each parameter up to 10.)

2.3. .PLOT

In most cases, tables of data are necessary. Tables, though, are usually not easy to interpret. A graph provides a picture of the circuit operation. Looking at large amounts of data in tables to determine the operation of a circuit is, at best, tedious. Although our imagination is good, it is difficult to see in our mind's eye how data in a table will look when they are graphed. A graph shows how a circuit behaves with changes in parameters. With a graph, we can see how a change in input will affect the output or other parameters of the circuit.

Making graphs is the function of .PLOT. When this command is used it will plot the values that were specified and calculated in the .OUT file for the circuit being examined. The graph is plotted in keyboard graphics, such as the asterisk, the plus sign, ×, and so on. If a graph intersects with another on the same plot, a capital X is printed at the intersection. Although this is not elegant, it is at least adequate. Another fact to be understood is that if the sweep function produced many values, the graph can be several pages long.

Keyboard graphics are the only type of graphics that the main program of PSpice has available. If higher-resolution graphics are desired or required, the graphics post-processor, PROBE, must be used. The format of the graph that is generated is that all of the output values are plotted along the y axis of the graph. The sweep values are plotted along the x axis of the graph. For each value calculated there is a point plotted on the graph. You can see that if many points are calculated, the graph can be very long. The syntax of this command is as follows:

```
.PLOT DC <output value> · · · <min range> <max range>
```

More than one .PLOT command is allowed in a netlist. If more than one parameter is to be plotted, a separate .PLOT command is not required for each parameter but may make the graphs more readable.

There is an additional parameter that may be used with .PLOT. We can specify a minimum and a maximum range for the plot. This means that you can specify the range of values to be plotted on the output axis of the graph. Note that only the output axis may be specified for its range; the input axis was specified in the sweep command. If you do not specify the output range, PSpice will set the range for you. The range that is set by PSpice will contain all of the output values. Thus if you are only interested in a specific small area of the total range of output, you would specify the range in the .PLOT command.

Let's add the .PLOT command to the bridge circuit and observe the graph of the change in voltage across the nodes (3,4) and the current in R_6 for the circuit. The type of sweep used in this netlist is linear, so we do not need to use the LIN statement since it is the default sweep. We can add the needed functions to the circuit by adding the necessary lines to the existing netlist.

```
*RESISTIVE BRIDGE DEMO CIRCUIT
VIN 1 0 10VOLTS
R1 1 2 10OHMS
R2 2 3 25OHMS
R3 3 0 10OHMS
R4 2 4 20OHMS
R5 4 0 50OHMS
R6 3 4 100OHMS
.DC VIN 0 10 .25
.PRINT DC V(3,0) V(4,0) I(R6) V(3,4)
```

```
.PLOT DC V(3,4)     ;graph of voltage across R6
.PLOT DC I(R6)      ;graph of current  in R6
.END
```

Notice that two plot functions were used in this netlist. This is not mandatory, but will make the graphs easier to read.

It is interesting to see what happens when we apply the decade and octave sweep to the bridge circuit. These are nonlinear sweeps and the curves produced by these sweeps will also be nonlinear, in this case logarithmic. The netlists for these sweeps are:

Decade sweep for the resistor bridge:

```
*RESISTIVE BRIDGE DEMO CIRCUIT
VIN 1 0 10VOLTS
R1 1 2 10OHMS
R2 2 3 25OHMS
R3 3 0 10OHMS
R4 2 4 20OHMS
R5 4 0 50OHMS
R6 3 4 100OHMS
.DC DEC VIN 0.1 10 25
.PRINT DC V(3,0) V(4,0) I(R6) V(3,4)
.PLOT DC V(3,4)
.PLOT DC I(R6)
.END
```

Octave sweep for the resistor bridge:

```
RESISTIVE BRIDGE DEMO CIRCUIT
VIN 1 0 10VOLTS
R1 1 2 10OHMS
R2 2 3 25OHMS
R3 3 0 10OHMS
R4 2 4 20OHMS
R5 4 0 50OHMS
R6 3 4 100OHMS
.DC OCT VIN 0.1 10 5
.PRINT DC V(3,0) V(4,0) I(R6) V(3,4)
.PLOT DC V(3,4)
.PLOT DC I(R6)
.END
```

Note that the starting value for the sweeps is not zero. The outputs of the three sweeps are shown in Analyses 2-1 to 2-3.

******* 01/04/80 ******* Evaluation PSpice (January 1990) ******* 14:49:13 *******

*RESISTIVE BRIDGE ANALYSIS 2-1

****   CIRCUIT DESCRIPTION

****

```
VIN 1 0 10
R1 1 2 10
R2 2 3 25
R3 3 0 10
R4 2 4 20
R5 4 0 50
R6 3 4 100
.DC VIN 0 10 0.5
.PRINT DC V(3) V(4) I(R6) V(3,4)
.PLOT DC V(3,4)
.PLOT DC I(R6)
.END
```

****   DC TRANSFER CURVES   TEMPERATURE =   27.000 DEG C

VIN	V(3)	V(4)	I(R6)	V(3,4)
0.000E+00	0.000E+00	0.000E+00	0.000E+00	0.000E+00
5.000E-01	1.077E-01	2.299E-01	-1.222E-03	-1.222E-01
1.000E+00	2.154E-01	4.598E-01	-2.445E-03	-2.445E-01
1.500E+00	3.231E-01	6.898E-01	-3.667E-03	-3.667E-01
2.000E+00	4.307E-01	9.197E-01	-4.889E-03	-4.889E-01
2.500E+00	5.384E-01	1.150E+00	-6.112E-03	-6.112E-01
3.000E+00	6.461E-01	1.380E+00	-7.334E-03	-7.334E-01
3.500E+00	7.538E-01	1.609E+00	-8.556E-03	-8.556E-01
4.000E+00	8.615E-01	1.839E+00	-9.779E-03	-9.779E-01
4.500E+00	9.692E-01	2.069E+00	-1.100E-02	-1.100E+00
5.000E+00	1.077E+00	2.299E+00	-1.222E-02	-1.222E+00
5.500E+00	1.185E+00	2.529E+00	-1.345E-02	-1.345E+00
6.000E+00	1.292E+00	2.759E+00	-1.467E-02	-1.467E+00
6.500E+00	1.400E+00	2.989E+00	-1.589E-02	-1.589E+00
7.000E+00	1.508E+00	3.219E+00	-1.711E-02	-1.711E+00
7.500E+00	1.615E+00	3.449E+00	-1.834E-02	-1.834E+00
8.000E+00	1.723E+00	3.679E+00	-1.956E-02	-1.956E+00
8.500E+00	1.831E+00	3.909E+00	-2.078E-02	-2.078E+00
9.000E+00	1.938E+00	4.139E+00	-2.200E-02	-2.200E+00
9.500E+00	2.046E+00	4.368E+00	-2.322E-02	-2.322E+00
1.000E+01	2.154E+00	4.598E+00	-2.445E-02	-2.445E+00

Analysis 2-1

```
******* 01/04/80 ******* Evaluation PSpice (January 1990) ******* 14:49:21
```

*RESISTIVE BRIDGE ANALYSIS 2-2

**** 　　　CIRCUIT DESCRIPTION

****

```
VIN 1 0 10
R1 1 2 10
R2 2 3 25
R3 3 0 10
R4 2 4 20
R5 4 0 50
R6 3 4 100
.DC DEC VIN 0.1 10 10
.PRINT DC V(3) V(4) I(R6) V(3,4)
.PLOT DC V(3,4)
.PLOT DC I(R6)
.options nopage
.END
```

**** 　　　DC TRANSFER CURVES 　　　　　　　TEMPERATURE = 　27.000 DEG C

VIN	V(3)	V(4)	I(R6)	V(3,4)
1.000E-01	2.154E-02	4.598E-02	-2.445E-04	-2.445E-02
1.259E-01	2.711E-02	5.789E-02	-3.078E-04	-3.078E-02
1.585E-01	3.413E-02	7.288E-02	-3.875E-04	-3.875E-02
1.995E-01	4.297E-02	9.175E-02	-4.878E-04	-4.878E-02
2.512E-01	5.410E-02	1.155E-01	-6.141E-04	-6.141E-02
3.162E-01	6.810E-02	1.454E-01	-7.731E-04	-7.731E-02
3.981E-01	8.574E-02	1.831E-01	-9.733E-04	-9.733E-02
5.012E-01	1.079E-01	2.305E-01	-1.225E-03	-1.225E-01
6.310E-01	1.359E-01	2.901E-01	-1.543E-03	-1.543E-01
7.943E-01	1.711E-01	3.653E-01	-1.942E-03	-1.942E-01
1.000E+00	2.154E-01	4.598E-01	-2.445E-03	-2.445E-01
1.259E+00	2.711E-01	5.789E-01	-3.078E-03	-3.078E-01
1.585E+00	3.413E-01	7.288E-01	-3.875E-03	-3.875E-01
1.995E+00	4.297E-01	9.175E-01	-4.878E-03	-4.878E-01
2.512E+00	5.410E-01	1.155E+00	-6.141E-03	-6.141E-01
3.162E+00	6.810E-01	1.454E+00	-7.731E-03	-7.731E-01
3.981E+00	8.574E-01	1.831E+00	-9.733E-03	-9.733E-01
5.012E+00	1.079E+00	2.305E+00	-1.225E-02	-1.225E+00
6.310E+00	1.359E+00	2.901E+00	-1.543E-02	-1.543E+00
7.943E+00	1.711E+00	3.653E+00	-1.942E-02	-1.942E+00
1.000E+01	2.154E+00	4.598E+00	-2.445E-02	-2.445E+00

Analysis 2-2

```
 ****          DC TRANSFER CURVES                    TEMPERATURE =      27.000 DEG C
    LEGEND:
 *: V(3,4)
    VIN           V(3,4)
 (*)----------     -3.0000E+00   -2.0000E+00   -1.0000E+00    2.2204E-16   1.0000E+00

    0.000E+00  0.000E+00 . - - - - - - - - - - - - - - - - - - * - - - -
    5.000E-01 -1.222E-01 .            .            .          *  .            .
    1.000E+00 -2.445E-01 .            .            .         *   .            .
    1.500E+00 -3.667E-01 .            .            .        *    .            .
    2.000E+00 -4.889E-01 .            .            .      *      .            .
    2.500E+00 -6.112E-01 .            .            .    *        .            .
    3.000E+00 -7.334E-01 .            .            .  *          .            .
    3.500E+00 -8.556E-01 .            .            *            .            .
    4.000E+00 -9.779E-01 .            .          *  .            .            .
    4.500E+00 -1.100E+00 .            .        * .               .            .
    5.000E+00 -1.222E+00 .            .       *    .            .            .
    5.500E+00 -1.345E+00 .            .      *     .            .            .
    6.000E+00 -1.467E+00 .            .    *       .            .            .
    6.500E+00 -1.589E+00 .            .  *         .            .            .
    7.000E+00 -1.711E+00 .            . *          .            .            .
    7.500E+00 -1.834E+00 .          * .            .            .            .
    8.000E+00 -1.956E+00 .         .*              .            .            .
    8.500E+00 -2.078E+00 .       *  .              .            .            .
    9.000E+00 -2.200E+00 .     *    .              .            .            .
    9.500E+00 -2.322E+00 .    *     .              .            .            .
    1.000E+01 -2.445E+00 .   *      .              .            .            .
```

```
 ****          DC TRANSFER CURVES                    TEMPERATURE =      27.000 DEG C
    VIN           I(R6)
 (*)----------     -3.0000E-02   -2.0000E-02   -1.0000E-02    1.7347E-18   1.0000E-02

    0.000E+00  0.000E+00 . - - - - - - - - - - - - - - - - - - * - - - -
    5.000E-01 -1.222E-03 .            .            .          *  .            .
    1.000E+00 -2.445E-03 .            .            .         *   .            .
    1.500E+00 -3.667E-03 .            .            .        *    .            .
    2.000E+00 -4.889E-03 .            .            .      *      .            .
    2.500E+00 -6.112E-03 .            .            .    *        .            .
    3.000E+00 -7.334E-03 .            .            .  *          .            .
    3.500E+00 -8.556E-03 .            .            *            .            .
    4.000E+00 -9.779E-03 .            .          *  .            .            .
    4.500E+00 -1.100E-02 .            .        * .               .            .
    5.000E+00 -1.222E-02 .            .       *    .            .            .
    5.500E+00 -1.345E-02 .            .      *     .            .            .
    6.000E+00 -1.467E-02 .            .    *       .            .            .
    6.500E+00 -1.589E-02 .            .  *         .            .            .
    7.000E+00 -1.711E-02 .            . *          .            .            .
    7.500E+00 -1.834E-02 .          * .            .            .            .
    8.000E+00 -1.956E-02 .         .*              .            .            .
    8.500E+00 -2.078E-02 .       *  .              .            .            .
    9.000E+00 -2.200E-02 .     *    .              .            .            .
    9.500E+00 -2.322E-02 .    *     .              .            .            .
    1.000E+01 -2.445E-02 .   *      .              .            .            .
```

```
    JOB CONCLUDED
    TOTAL JOB TIME                 7.36
```

Analysis 2-2 (cont'd)

```
****      DC TRANSFER CURVES                    TEMPERATURE =    27.000 DEG C

LEGEND:

*: V(3,4)
   VIN        V(3,4)
(*)----------    -3.0000E+00  -2.0000E+00  -1.0000E+00  2.2204E-16   1.0000E+00

    1.000E-01 -2.445E-02 . _ _ _ _ _ _ _ _ _ _ _ _ _ _ _ _ _ _ *_ _ _ _ _ .
    1.259E-01 -3.078E-02 .                .            .          *        .
    1.585E-01 -3.875E-02 .                .            .         *.        .
    1.995E-01 -4.878E-02 .                .            .         *.        .
    2.512E-01 -6.141E-02 .                .            .         *.        .
    3.162E-01 -7.731E-02 .                .            .         *.        .
    3.981E-01 -9.733E-02 .                .            .         *.        .
    5.012E-01 -1.225E-01 .                .            .        * .        .
    6.310E-01 -1.543E-01 .                .            .       *  .        .
    7.943E-01 -1.942E-01 .                .            .       *  .        .
    1.000E+00 -2.445E-01 .                .            .       *  .        .
    1.259E+00 -3.078E-01 .                .            .      *   .        .
    1.585E+00 -3.875E-01 .                .            .     *    .        .
    1.995E+00 -4.878E-01 .                .            .     *    .        .
    2.512E+00 -6.141E-01 .                .            .   *      .        .
    3.162E+00 -7.731E-01 .                .            . *        .        .
    3.981E+00 -9.733E-01 .                .          * .          .        .
    5.012E+00 -1.225E+00 .                .       *    .          .        .
    6.310E+00 -1.543E+00 .                .   *        .          .        .
    7.943E+00 -1.942E+00 .             .* .            .          .        .
    1.000E+01 -2.445E+00 .         *      .            .          .        .
                         . _ _ _ _ _ _ _ _ _ _ _ _ _ _ _ _ _ _ _ _ _ _ _ _ .

****      DC TRANSFER CURVES                    TEMPERATURE =    27.000 DEG C

   VIN        I(R6)
(*)----------    -3.0000E-02  -2.0000E-02  -1.0000E-02  1.7347E-18   1.0000E-02

    1.000E-01 -2.445E-04 . _ _ _ _ _ _ _ _ _ _ _ _ _ _ _ _ _ _ *_ _ _ _ _ .
    1.259E-01 -3.078E-04 .                .            .          *        .
    1.585E-01 -3.875E-04 .                .            .         *.        .
    1.995E-01 -4.878E-04 .                .            .         *.        .
    2.512E-01 -6.141E-04 .                .            .         *.        .
    3.162E-01 -7.731E-04 .                .            .         *.        .
    3.981E-01 -9.733E-04 .                .            .         *.        .
    5.012E-01 -1.225E-03 .                .            .        * .        .
    6.310E-01 -1.543E-03 .                .            .        * .        .
    7.943E-01 -1.942E-03 .                .            .       *  .        .
    1.000E+00 -2.445E-03 .                .            .       *  .        .
    1.259E+00 -3.078E-03 .                .            .      *   .        .
    1.585E+00 -3.875E-03 .                .            .     *    .        .
    1.995E+00 -4.878E-03 .                .            .    *     .        .
    2.512E+00 -6.141E-03 .                .            .  *       .        .
    3.162E+00 -7.731E-03 .                .            .*         .        .
    3.981E+00 -9.733E-03 .                .          * .          .        .
    5.012E+00 -1.225E-02 .                .       *    .          .        .
    6.310E+00 -1.543E-02 .                .   *        .          .        .
    7.943E+00 -1.942E-02 .             .* .            .          .        .
    1.000E+01 -2.445E-02 .         *      .            .          .        .
                         . _ _ _ _ _ _ _ _ _ _ _ _ _ _ _ _ _ _ _ _ _ _ _ _ .

          JOB CONCLUDED
          TOTAL JOB TIME            5.71
```

Analysis 2-2 (cont'd)

```
******* 01/04/80 ******* Evaluation PSpice (January 1990) ******* 14:49:28

*RESISTIVE BRIDGE ANALYSIS 2-3

****     CIRCUIT DESCRIPTION

*******************************************************************************

VIN 1 0 10
R1 1 2 10
R2 2 3 25
R3 3 0 10
R4 2 4 20
R5 4 0 50
R6 3 4 100
.DC OCT VIN 0.1 10 2
.PRINT DC V(3) V(4) I(R6) V(3,4)
.PLOT DC V(3,4)
.PLOT DC I(R6)
.options nopage
.END
```

```
****     DC TRANSFER CURVES          TEMPERATURE =   27.000 DEG C

     VIN          V(3)         V(4)         I(R6)        V(3,4)

    1.000E-01    2.154E-02    4.598E-02   -2.445E-04   -2.445E-02
    1.414E-01    3.046E-02    6.503E-02   -3.457E-04   -3.457E-02
    2.000E-01    4.307E-02    9.197E-02   -4.889E-04   -4.889E-02
    2.828E-01    6.091E-02    1.301E-01   -6.915E-04   -6.915E-02
    4.000E-01    8.615E-02    1.839E-01   -9.779E-04   -9.779E-02
    5.657E-01    1.218E-01    2.601E-01   -1.383E-03   -1.383E-01
    8.000E-01    1.723E-01    3.679E-01   -1.956E-03   -1.956E-01
    1.131E+00    2.437E-01    5.202E-01   -2.766E-03   -2.766E-01
    1.600E+00    3.446E-01    7.357E-01   -3.912E-03   -3.912E-01
    2.263E+00    4.873E-01    1.040E+00   -5.532E-03   -5.532E-01
    3.200E+00    6.892E-01    1.471E+00   -7.823E-03   -7.823E-01
    4.525E+00    9.746E-01    2.081E+00   -1.106E-02   -1.106E+00
    6.400E+00    1.378E+00    2.943E+00   -1.565E-02   -1.565E+00
    9.051E+00    1.949E+00    4.162E+00   -2.213E-02   -2.213E+00
    1.280E+01    2.757E+00    5.886E+00   -3.129E-02   -3.129E+00
```

Analysis 2-3

```
****      DC TRANSFER CURVES                    TEMPERATURE =    27.000 DEG C

LEGEND:

*: V(3,4)
  VIN         V(3,4)
(*)----------     -4.0000E+00  -3.0000E+00  -2.0000E+00  -1.0000E+00   0.0000E+00
                  - - - - - - - - - - - - - - - - - - - - - - - - - -
 1.000E-01 -2.445E-02 .            .            .            .            *
 1.414E-01 -3.457E-02 .            .            .            .            *
 2.000E-01 -4.889E-02 .            .            .            .            *.
 2.828E-01 -6.915E-02 .            .            .            .            *.
 4.000E-01 -9.779E-02 .            .            .            .            *.
 5.657E-01 -1.383E-01 .            .            .            .           * .
 8.000E-01 -1.956E-01 .            .            .            .          *  .
 1.131E+00 -2.766E-01 .            .            .            .         *   .
 1.600E+00 -3.912E-01 .            .            .            .        *    .
 2.263E+00 -5.532E-01 .            .            .            .    *        .
 3.200E+00 -7.823E-01 .            .            .            .  *          ,
 4.525E+00 -1.106E+00 .            .            .         *.               .
 6.400E+00 -1.565E+00 .            .            .    *                     .
 9.051E+00 -2.213E+00 .            .       *    .                          .
 1.280E+01 -3.129E+00 .      *     .            .                          .
                  - - - - - - - - - - - - - - - - - - - - - - - - - -


****      DC TRANSFER CURVES                    TEMPERATURE =    27.000 DEG C

  VIN         I(R6)
(*)----------     -4.0000E-02  -3.0000E-02  -2.0000E-02  -1.0000E-02   0.0000E+00
                  - - - - - - - - - - - - - - - - - - - - - - - - - -
 1.000E-01 -2.445E-04 .            .            .            .            *
 1.414E-01 -3.457E-04 .            .            .            .            *
 2.000E-01 -4.889E-04 .            .            .            .            *.
 2.828E-01 -6.915E-04 .            .            .            .            *.
 4.000E-01 -9.779E-04 .            .            .            .            *.
 5.657E-01 -1.383E-03 .            .            .            .           * .
 8.000E-01 -1.956E-03 .            .            .            .          *  .
 1.131E+00 -2.766E-03 .            .            .            .         *   .
 1.600E+00 -3.912E-03 .            .            .            .        *    .
 2.263E+00 -5.532E-03 .            .            .            .    *        .
 3.200E+00 -7.823E-03 .            .            .            .  *          .
 4.525E+00 -1.106E-02 .            .            .         *.               .
 6.400E+00 -1.565E-02 .            .            .    *                     .
 9.051E+00 -2.213E-02 .            .       *    .                          .
 1.280E+01 -3.129E-02 .      *     .            .                          .
                  - - - - - - - - - - - - - - - - - - - - - - - - - -

          JOB CONCLUDED

          TOTAL JOB TIME        5.32
```

Analysis 2-3 (cont'd)

2.4. AC FUNCTIONS IN PSPICE

Any analysis program must be able to analyze both dc and ac circuits. To do this the program must also be able to handle inductance and capacitance as well as resistance. Inductance and capacitance also appear in dc circuits, but we have put off describing how to work with them until now. Both dc and ac handling of inductance and capacitance are the same.

2.5. HANDLING INDUCTANCE AND CAPACITANCE IN PSPICE

The resistances that we have been using so far present no problem. They are limiting conductors and therefore limit the current that can flow in the circuit. This is not the case for C and L. PSpice treats both of these devices as perfect. This produces problems that are specific to each of the devices. The perfect capacitor conducts no dc current, while the perfect inductor has zero resistance and can conduct enormous dc current. For this reason certain precautions must be taken with each of these devices. If they are not handled properly, errors will be reported by the program.

2.5.1. Using Capacitance in a Netlist

As stated before, all nodes must have a dc return to ground. The perfect capacitor does not conduct any dc current (it is an open circuit), so it does not meet this requirement. When handling capacitance we must provide a dc path to the ground node. The method used to meet this need is to place a large resistance in parallel with the capacitance. The size of this resistance is usually in the range of gigaohms. This is large enough that the operation of the circuit will not be measurably disturbed. PSpice does not care how much current is conducted, only that a dc path to the reference node exists. Figure 2-1 shows a simple two-capacitor circuit with a resistor to provide a path to ground for the capacitors. The netlist for this circuit is

```
*TWO CAPACITOR PROGRAM
VDC 1 0 10
C1  1 2 1U
C2  2 0 1U
R1  2 0 1G
.END
```

Simple analysis of Figure 2-1 shows that the resistor is across both of the capacitors. This is all that is needed for this circuit. If necessary, a resistance larger than this value can be used. If no resistor is used, a "NO DC PATH TO NODE" error will be the error reported in the output file. If a resistance is already part of the circuit being analyzed, and is across the capacitors in the circuit and providing a path to ground, there is no need for another resistor.

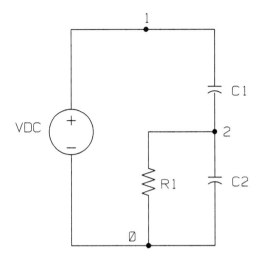

Figure 2-1

2.5.2. Handling Inductance

Inductances can also be a problem because PSpice sees a perfect inductor. The inductor has zero resistance and is treated as a perfect voltage source. For this reason some resistance must be included in series with the inductor or an error message will be generated. The size of the resistance to be included with the inductor is determined by the type of circuit being investigated. If the resistance of the inductor itself is large, this value can be used as the series resistor. If an inductor is part of a resonant circuit at a high frequency, resistance is very important, as it influences Q. A good value of resistance may be 0.001 Ω or less. In any case, the value of resistance to be used is determined by the circuit application.

Another point to be careful with when using an inductor is the voltage loop problem. If the circuit has an inductor directly across the source and there is no resistance in the inductor path, the program will abort and report an error. The error will be "VOLTAGE LOOP INVOLVING V(X)".

2.6. PSpice AC INPUT SOURCE SPECIFICATIONS

Until now we have not had a need to specify the type of input source to PSpice. The dc source is the default source in PSpice. Since we are now working with ac, we must know how to specify an ac source. We also need to know the differences between ac and dc source specification. Unlike dc applications, we must specify ac in the source netlist line. PSpice automatically determines that any source that is not specified as to its type is a dc source. PSpice cannot determine what is in our mind, only what it is told. So we must label an ac source as ac.

The form of the source statement for any circuit can include both dc and ac sources on the same line. The syntax for using dc and ac sources together is

```
V/I<name> <+node> <-node> DC <voltage/current value>
+ AC <voltage/current value> <magnitude> <phase>
```

When working with ac, we work not only with how much voltage there is, but also with the frequency and phase of the voltage. For this reason the ac voltage specification uses both voltage magnitude and phase. If the phase angle is not defined, PSpice assumes that the phase angle of the applied frequency is zero. The magnitude is the peak value of the applied voltage. PSpice uses only peak values for ac voltages. The phase is the offset value of the phase of the source relative to zero phase. In some cases this is not important since we assume that the input source is at zero phase, and this is the reference for all other phases in the circuit. PSpice can handle as many inputs of different magnitudes and phases as we need.

The ac input magnitude is not always the most important parameter. Often we are more interested in the gain of the circuit we are analyzing. Since the voltage or current levels in a ac circuit usually change in proportion to both the input magnitude and frequency, the input is usually specified as unity. This makes the output equal to the numerical gain of the circuit, but can also lead to very large output voltages for active device circuits. This is all right since the decibel gain of the circuit will not change.

There are some special considerations that we must know when we use the ac functions. We have both magnitude and phase for voltage and current in an ac analysis. The standard V or I is used for these sources, but there are some suffixes that may be added to V and I to print or plot what is needed. A list of the suffixes is given in Table 2-1. A brief discussion of these values is in order here, so we can use them properly.

TABLE 2-1

Suffix	Meaning
M (or no suffix)	Magnitude (default value)
DB	Magnitude in decibels
P	Phase
G	Group delay
R	Real part of a voltage or current
I	Imaginary part of a voltage or current

When used with either a voltage or a current, the M suffix indicates that we want to know the magnitude of the ac voltage or current. M is also the default value for voltage or current. If M is not included in the statement, we will automatically get the magnitude of the voltage or current. An example of the use of this suffix is

```
.PRINT AC V(3) V(5)
.PLOT AC VM(5) VP(5)
.PRINT AC IM(R6)
```

The DB suffix is used when we wish to know either the number of decibels for a given voltage referenced to zero or the number of decibels between two voltage levels. Decibels (dB) are calculated in PSpice by

$$dB = 20 \log \frac{V_o}{V_i}$$

An example of how to specify the decibel function is

```
DB(V(8)/V(5))
VDB(8)
```

where the first line indicates the decibel value of V_8 divided by V_5, and the second line is the decibel value relative to zero decibels. The P suffix gives the phase of the ac voltage referenced to zero phase.

```
VP(8)
VP(8) - VP(5)
```

The G suffix gives the group delay of a voltage over the x-axis values. The syntax for this function is

```
VG(8)
```

The R suffix gives the real component of a voltage or current that has both imaginary and real components. The syntax for this is

```
VR(8), IR(R8)
```

The I suffix gives the imaginary component of a voltage or current that has both imaginary and real components. The syntax for this is

```
VI(8), II(R8)
```

2.7. USING .PRINT AND .PLOT FOR AC FUNCTIONS

As in dc analysis, the output from the ac analysis may be printed or plotted using the .PRINT and .PLOT functions. The main difference is that the output is arranged by frequency rather than by voltage. A value is printed or plotted at the frequency at which the calculation was made. The syntax for these commands is

```
.PRINT AC <output value> · · ·
.PLOT AC <output value> · · ·
```

Once again, there may be as many output values as necessary. Each of the output values that are calculated becomes an entry in the column that is specified in the command. The order of the columns is the same as in the .DC function. The output values that may be either printed or plotted are node, device, and source voltages and currents, magnitude, real and imaginary voltage and current components, and phase angles.

2.8. THE .AC SWEEP FUNCTION

This is the ac counterpart to the .DC sweep function. It sweeps a range of frequencies that are set by you. There are differences between the .DC function and the .AC function and you must be aware of them. The frequencies that are specified in the sweep are applied to all of the ac sources in the circuit at the same time. It is not possible to perform two different ac sweeps in the same circuit. Thus a nested sweep is not possible either. The .AC function has three forms: (1) linear, (2) decade, and (3) octave.

There is no form of the LIST type of sweep for ac. The syntax for the three types of sweep is as follows:

1. *Linear sweep:*

```
.AC LIN <points value> <start value> <end value>
```

This defines a linear sweep of frequency with a specified number of equally spaced points to be calculated. The points' value specifies how many points are to be calculated for the entire sweep. The sweep starts at the specified value (this value may or may not be zero) and ends at the specified value. Note that the input source being swept is not named in the sweep specification. This is because all ac sources in a circuit are swept simultaneously through the same range.

This type of sweep is useful for many analyses. It is also useful for analyzing a circuit at a single frequency. We may need to know all the voltages and currents, with their magnitude, real and imaginary components, and the phase angles of the voltages and currents in the circuit.

Let us analyze a simple *RC* circuit using this type of sweep at a single frequency. We shall try to get all of the data we can from the circuit. The circuit is shown in Figure 2-2. The netlist for the circuit is

```
*SINGLE AC SWEEP POINT RC CIRCUIT ANALYSIS 2-4
VAC 1 0 AC 120Volts
R1 1 2 200
R2 2 0 400
C1 2 0 10U
.AC LIN 1 60 60
.PRINT AC VM(1,2) IM(R1) VP(1,2) VM(2) IM(R2) VP(2)
+        VR(1,2) VI(1,2) IR(R1) II(R1) VR(2) IR(R2)
+        II(R2) IR(C1) II(C1)
.END
```

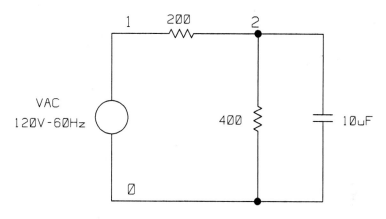

Figure 2-2

```
******* 10/11/90 ******* Evaluation PSpice (January 1990) ******* 07:32:52 *******

*SINGLE AC SWEEP POINT RC CIRCUIT ANALYSIS 2-4

****        CIRCUIT DESCRIPTION

***********************************************************************
VAC 1 0 AC 120Volts
R1 1 2 200
R2 2 0 400
C1 2 0 10U
.AC LIN 1 60 60
.PRINT AC VM(1,2) IM(R1) VP(1,2) VM(2) IM(R2) VP(2
+       VR(1,2) VI(1,2) IR(R1) II(R1) VR(2) IR(R2)
+       II(R2) IR(C1) II(C1)
.END

******* 10/11/90 ******* Evaluation PSpice (January 1990) ******* 07:32:52 *******

*SINGLE AC SWEEP POINT RC CIRCUIT ANALYSIS 2-4

****        SMALL SIGNAL BIAS SOLUTION       TEMPERATURE =   27.000 DEG C

***********************************************************************
NODE   VOLTAGE       NODE   VOLTAGE       NODE   VOLTAGE       NODE   VOLTAGE

(   1)   0.0000  (    2)    0.0000

    VOLTAGE SOURCE CURRENTS
    NAME           CURRENT

    VAC            0.000E+00

    TOTAL POWER DISSIPATION   0.00E+00  WATTS
```

Analysis 2-4

```
******* 10/11/90 ******* Evaluation PSpice (January 1990) ******* 07:32:52 *******

*SINGLE AC SWEEP POINT RC CIRCUIT ANALYSIS 2-4

****     AC ANALYSIS                      TEMPERATURE =   27.000 DEG C

**********************************************************************

  FREQ      VM(1,2)     IM(R1)     VP(1,2)     VM(2)      IM(R2)


 6.000E+01  6.467E+01  3.233E-01  2.976E+01  7.148E+01  1.787E-01

******* 10/11/90 ******* Evaluation PSpice (January 1990) ******* 07:32:52 *******

*SINGLE AC SWEEP POINT RC CIRCUIT ANALYSIS 2-4

****     AC ANALYSIS                      TEMPERATURE =   27.000 DEG C

**********************************************************************

  FREQ      VP(2)      VR(1,2)     VI(1,2)     IR(R1)     II(R1)

 6.000E+01 -2.669E+01  5.614E+01  3.210E+01  2.807E-01  1.605E-01

******* 10/11/90 ******* Evaluation PSpice (January 1990) ******* 07:32:52 *******

*SINGLE AC SWEEP POINT RC CIRCUIT ANALYSIS 2-4

****     AC ANALYSIS                      TEMPERATURE =   27.000 DEG C

**********************************************************************

  FREQ      VR(2)      IR(R2)     II(R2)     IR(C1)     II(C1)

 6.000E+01  6.386E+01  1.597E-01 -8.025E-02  1.210E-01  2.408E-01

            JOB CONCLUDED

            TOTAL JOB TIME            4.12
```

Analysis 2-4 (cont'd)

When this analysis is run, the output file will contain the node voltages and device currents shown above. The table will be in the order shown. The output file is shown in Analysis 2-4.

A brief explanation of the output file of the single-point analysis is in order here. In all cases, PSpice calculates a dc bias point for any circuit even if the dc voltages are zero. The output file shows that all dc voltages and currents are zero. The ac analysis that was done is stored in the output file, with a maximum of five entries (columns) per line. For this circuit, there are 15 values calculated. There are three sets of analyses presented. The analyses are in the order of the .PRINT command that we used. The .PRINT command was necessary in order for the calculated values to be stored.

If you were to print this information out to a printer, you find that you have about five pages, with only one group of data per page. This is somewhat wasteful. You can suppress the paging of the printer by using the command

```
.OPTIONS NOPAGE
```

This command, which can go anywhere between the title line and the ending line, allows us to change a number of parameters of the program. More will be said about this command as needed. It is enough for the moment to say that it removes the form feed character from the output file.

2. *Decade sweep:*

```
.AC DEC <points/decade value> <start value> <end value>
```

The decade sweep is a logarithmic sweep in decades (a 10:1 ratio) just like the dc decade sweep. You must specify the number of decades to be swept by this function. It is not mandatory to specify only in decades. You can also specify a sweep with two values that constitute all or part of single or multiple decades. 20 to 800 kilohertz (kHz) or 2 to 8 kHz are acceptable values. The range of frequencies that can be swept is quite large, beyond 10 GHz.

The points value for this sweep indicates the number of points that are calculated for each decade of sweep. Once again, zero is not acceptable in either the decade or octave sweep. Note also that the source is not named in the sweep command line. What is important for this sweep is the frequency range, and this range will be the same for all sources in the circuit.

3. *Octave sweep:*

```
.AC OCT <points/octave value> <start value> <end value>
```

This is also a logarithmic sweep. In this sweep, the number of points that is specified is the number of points that will be calculated for each octave change in frequency. (An octave is a 2-to-1 change in frequency, such as 100 Hz to 200 Hz, 200 Hz to 400 Hz, etc.) In the octave sweep you must choose the number of points per octave carefully since the frequency change is only 2 to 1. If there are 50 octaves to be calculated and you choose 20 points per octave, you will calculate 1001 points!

Let us analyze a simple *LC* filter circuit to see how the ac sweep works. The circuit is an *m*-derived (M = 0.6) filter, with a −3-dB frequency of 400 kHz and a maximum attenuation frequency of 600 kHz. The input and output impedance of the filter is 50 Ω. The circuit for this filter is shown in Figure 2-3. The netlist for the filter is

```
*M-DERIVED FILTER DEMO
VIN 1 0 AC 1
R1 1 2 50
L1 2 3 19.1U
C1 2 0 .0039U
C2 2 3 .0033U
C3 3 0 .0039U
```

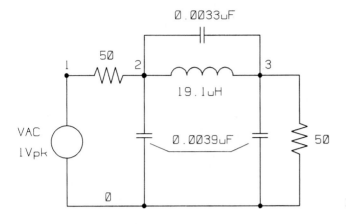

Figure 2-3

```
RL 3 0 50
.AC DEC 10 1E4 1E7
.PRINT AC V(3) IP(R1)
.PLOT AC V(3) VP(3)
.END
```

The print and plot output of this netlist are shown in Analysis 2-5.

     An important note here regards the number of points for the decade sweep. Notice that only 10 points per decade are plotted. The more points calculated, the longer the curve that is plotted. It is easy to get a graph several pages long. This is usually of no benefit since it makes the graph hard to read. You will want to stay around this number for keyboard graphics plotting.

******* 01/04/80 ******* Evaluation PSpice (January 1990) ******* 17:06:34 *******

*M-DERIVED LOW FREQUENCY RF FILTER DEMO ANALYSIS 2-5

****   CIRCUIT DESCRIPTION

****

```
VIN 1 0 AC 1
RIN 1 2 50
C1 2 0 .0039U
C2 2 3 .0033U
C3 3 0 .0039U
L1 2 3 19.1U
RL 3 0 50
.PRINT AC V(3)
.PLOT AC V(3)
.AC DEC 10 1E4 1E7
.END
```

Analysis 2-5

```
******* 01/04/80 ******* Evaluation PSpice (January 1990) ******* 17:06:34 *******

*M-DERIVED LOW FREQUENCY RF FILTER DEMO ANALYSIS 2-5

****      SMALL SIGNAL BIAS SOLUTION           TEMPERATURE =   27.000 DEG C

*********************************************************************************


  NODE   VOLTAGE      NODE   VOLTAGE      NODE   VOLTAGE      NODE   VOLTAGE

(    1)    0.0000  (    2)    0.0000  (    3)    0.0000



      VOLTAGE SOURCE CURRENTS
      NAME         CURRENT

      VIN          0.000E+00

      TOTAL POWER DISSIPATION   0.00E+00   WATTS



  ******* 01/04/80 ******* Evaluation PSpice (January 1990) ******* 17:06:34 *******

  *M-DERIVED LOW FREQUENCY RF FILTER DEMO ANALYSIS 2-5

  ****      AC ANALYSIS                        TEMPERATURE =   27.000 DEG C

  *********************************************************************************


   FREQ        V(3)

    1.000E+04    5.000E-01
    1.259E+04    5.000E-01
    1.585E+04    5.000E-01
    1.995E+04    5.000E-01
    2.512E+04    5.000E-01
    3.162E+04    5.000E-01
    3.981E+04    5.000E-01
    5.012E+04    5.000E-01
    6.310E+04    5.000E-01
    7.943E+04    5.000E-01
    1.000E+05    5.000E-01
    1.259E+05    5.000E-01
    1.585E+05    4.999E-01
    1.995E+05    4.997E-01
    2.512E+05    4.983E-01
```

Analysis 2-5 (cont'd)

```
3.162E+05    4.909E-01
3.981E+05    4.492E-01
5.012E+05    2.656E-01
6.310E+05    3.900E-03
7.943E+05    1.150E-01
1.000E+06    1.474E-01
1.259E+06    1.459E-01
1.585E+06    1.314E-01
1.995E+06    1.128E-01
2.512E+06    9.399E-02
3.162E+06    7.697E-02
3.981E+06    6.233E-02
5.012E+06    5.012E-02
6.310E+06    4.012E-02
7.943E+06    3.202E-02
1.000E+07    2.552E-02
```

```
******* 01/04/80 ******* Evaluation PSpice (January 1990) ******* 17:06:34 *******


*M-DERIVED LOW FREQUENCY RF FILTER DEMO ANALYSIS 2-5


****     AC ANALYSIS                      TEMPERATURE =    27.000 DEG C


*******************************************************************************



    FREQ       V(3)


(*)----------    1.0000E-03   1.0000E-02   1.0000E-01   1.0000E+00   1.0000E+01
            - - - - - - - - - - - - - - - - - - - - - - - - - - - - - - - -
 1.000E+04  5.000E-01 .                .            .          *  .              .
 1.259E+04  5.000E-01 .                .            .          *  .              .
 1.585E+04  5.000E-01 .                .            .          *  .              .
 1.995E+04  5.000E-01 .                .            .          *  .              .
 2.512E+04  5.000E-01 .                .            .          *  .              .
 3.162E+04  5.000E-01 .                .            .          *  .              .
 3.981E+04  5.000E-01 .                .            .          *  .              .
 5.012E+04  5.000E-01 .                .            .          *  .              .
 6.310E+04  5.000E-01 .                .            .          *  .              .
 7.943E+04  5.000E-01 .                .            .          *  .              .
 1.000E+05  5.000E-01 .                .            .          *  .              .
 1.259E+05  5.000E-01 .                .            .          *  .              .
 1.585E+05  4.999E-01 .                .            .          *  .              .
 1.995E+05  4.997E-01 .                .            .          *  .              .
 2.512E+05  4.983E-01 .                .            .         *   .              .
 3.162E+05  4.909E-01 .                .            .        *    .              .
 3.981E+05  4.492E-01 .                .            .        *    .              .
 5.012E+05  2.656E-01 .                .            .   *         .              .
 6.310E+05  3.900E-03 .         *      .            .            .              .
 7.943E+05  1.150E-01 .                .          .*            .              .
 1.000E+06  1.474E-01 .                .          . *           .              .
 1.259E+06  1.459E-01 .                .          . *           .              .
 1.585E+06  1.314E-01 .                .          .*            .              .
 1.995E+06  1.128E-01 .                .         .*             .              .
```

Analysis 2-5 (cont'd)

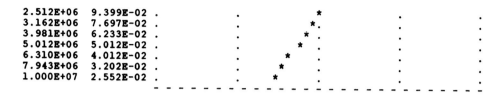

```
2.512E+06  9.399E-02 .              .              *          .              .
3.162E+06  7.697E-02 .              .            *.           .              .
3.981E+06  6.233E-02 .              .          *  .           .              .
5.012E+06  5.012E-02 .              .         *   .           .              .
6.310E+06  4.012E-02 .              .       *     .           .              .
7.943E+06  3.202E-02 .              .     *       .           .              .
1.000E+07  2.552E-02 .              .    *        .           .              .
                     - - - - - - - - - - - - - - - - - - - - - - - - - - - - -
```

       JOB CONCLUDED

       TOTAL JOB TIME   8.68

Analysis 2-5 (cont'd)

2.9. USING PROBE FOR HIGH-RESOLUTION GRAPHICS

It should be obvious that the curves for this filter are more interesting than the curves of a purely resistive circuit that is linear. Also obvious is that the .PLOT function leaves a lot to be desired in the form of the printed output of these filters. PSpice has a high-resolution graphics postprocessor called PROBE that is similar to having an oscilloscope with which to look at the circuit responses. This graphics postprocessor will produce all of the output curves of any circuit that we design. PROBE also has a menu that will allow us to alter the PROBE screen. From here on we will be working with .PROBE almost exclusively. PROBE can be executed from the keyboard; this is handy if you make a mistake and exit the program before you are ready to, or you need to quit without analyzing the output.

It is important to point out that the version of PSpice you are using will make a difference in the type of PROBE menu you will see. If you are using the student version, PROBE will run automatically when the .PROBE command is in the netlist and you will see a PROBE menu like the one shown below.

Advanced version PROBE menus   If you are using the advanced version or the evaluation version, you may see one of two menus, depending on the version you have. Both the advanced version and the evaluation version run in a shell, and PROBE runs automatically in the older version of both, but not in the latest version. In the shell, all of the options are on pull-down menus. Select the menu option by typing the first letter of the menu name.

When PROBE is run, the older advanced version menu is the same as the student version, but the latest version has a new menu. The latest version as of this writing is shown below. In the latest version, the selection of the desired function is done by pressing the first letter of the function, or pressing the return key for the value highlighted in reverse video. All the same functions are available in all versions. To make PROBE run automatically, use the PROBE menu and select the automatic function.

All of the netlists in this book have been run on the student version of PSpice.

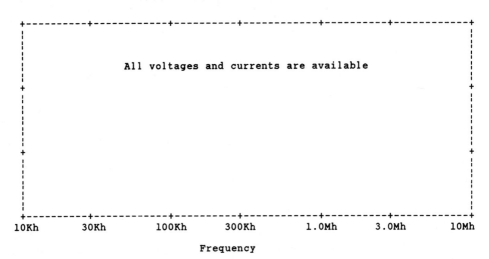

All voltages and currents are available

Frequency

Exit Add_trace X_axis Y_axis Plot_control Display_control Macros
Hard_copy

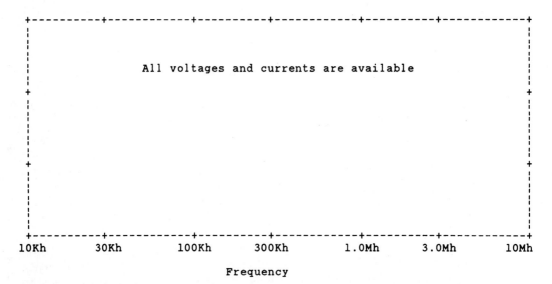

All voltages and currents are available

Frequency

0)Exit 1)Add Trace 3)X Axis 4)Y Axis 5)Add Plot 8)Hard Copy
9)Suppress Symbols : 1

The differences between the more advanced versions and the student version will be
pointed out as they are encountered. Also important here is the fact that none of the
data generated is kept in RAM but is always written to disk. If you do not have a hard
disk or high-density disk drives, you may run into available storage capacity problems.

With a 360K disk drive, if you have the minimum number of files needed to run PSpice on the disk, you will have about 70K to 90K of space left on the disk. This can cause PSpice to abort in some of the longer analyses to be done in this book. All is not lost— the data generated prior to the abort are still available and can be plotted.

You must also have the file PROBE.DEV properly set up for your printer and graphics system. This is necessary regardless of the version you are using. This is not difficult; we merely edit the file and add the correct devices to it. The PROBE.DEV file is on the disk with the PROBE.EXE file. As shipped, the devices in the file are

```
DISPLAY = TEXT
HARD-COPY = PRN, TEXT
```

These two parameters need to be changed to the type of printer and display you are using. PSpice has a number of printers, plotters, and displays from which to choose. To see a list of the devices, you should edit the PROBE.DEV file and change the display to something such as XXXX (DISPLAY = XXXX). PSpice will not recognize this as a valid display and will show a list of the display types that are available. Write down the one corresponding to your display type. Do a similar thing for the hard-copy option. Then edit the file to contain the printer and CRT you are using. You will then have the high-resolution graphics needed for the analyses to follow.

The general syntax for PROBE in a netlist is

```
.PROBE <output value> · · ·
```

More than one output value may be calculated. However, if you specify output values, PROBE will calculate only the values listed. The best use of the .PROBE command is to omit all of the output values. This forces PROBE to calculate all parameters and store them on the disk. The output parameters are stored in a file called PROBE.DAT for student PSpice and under the circuit name with the extension .DAT for the other versions. If you should run into storage space problems in a long analysis, you can use the output parameter designation to calculate only the parameters needed, saving disk space.

When you do not specify output parameters, you can plot all the parameters of the circuit. You may also have more than one parameter (plot) on the same graph. This can be done without rerunning the netlist, as would be the case with .PLOT. If one of the graphs proves not to be of much value, you can remove it and allow the others to remain. You can also remove all of the graphs and start over without rerunning the program. If you have an area of the graph that is of special importance, you can rescale the axes to show just the needed portion. You can also add separate graphs with their own y axes. PROBE also has a hard-copy option that allows you to print out the data shown on the screen. There are several options for the hard copy. It may be one page long, two pages long, or as long as you need it to be. The options are shown on a menu line below the graph.

```
                        VDDDDDDD7
                        : Probe :
                        SDDDDDDD=
              Graphics Post-Processor for PSpice
                  Version 1.13 - December 1987
         (C) Copyright 1985, 1986, 1987 by MicroSim Corporation
                        DDDDDDDDD
      DEMO VERSION: Copying of this program is welcomed and encouraged.



   Circuit:  *M-DERIVED LOW FREQUENCY RF FILTER DEMO
   Date/Time run:  10/08/90  09:35:09                  Temperature:    27.0
```

```
   0)Exit Program  1)DC Sweep  2)AC Sweep  : 1
```

Figure 2-4

2.10. MENU FUNCTIONS OF .PROBE

When the netlist for the filter is run, PROBE runs automatically (if you are using the student version) because the command line .PROBE is in the circuit netlist. Since we are doing only an ac analysis for frequency and output amplitude, PROBE goes directly to its graphics screen. This screen has the frequencies we specified for the netlist along the x axis, and a blank y axis.

It is possible to run a ac and dc analysis in the same netlist. If we had run both a dc and ac analysis, a menu screen would be shown first. This would let us choose the type of analysis we wish to examine first. A menu screen is shown in Figure 2-4. After the program runs, the screen you will see is one of the two shown earlier (if there are no errors in the netlist). If you are running the latest version, you will need to select the PROBE menu to run PROBE.

If you have the PROBE screen before you now, we can start with what the basic menu under the startup screen shows. Then we will see what some of the submenus will show. The screen indicates that all the currents and voltages for this circuit are available. The PROBE menu shown is for the student version, but all the functions are available on the advanced versions as well. The explanations apply to both menus. You are asked to make a choice here; the choices are as shown for the first menu of PROBE.

```
   BASIC GRAPHICS MENU: (student version)
      0)EXIT 1)ADD TRACE 3)X AXIS 4)Y AXIS 5)PLOT CONTROL
      8)HARD COPY 9)SUPPRESS SYMBOLS : 1
```

The numbers shown are for the number keys above the keyboard and all of the number keys are used. (Some are used with submenus.)

0)   EXIT   In all cases zero is the escape from the function you choose. You must be careful when using this key. If you accidentally press zero too many times, you will return to the startup screen or exit the program altogether. The exit function means what it says. It will exit either the function you have selected or the entire program. It really does not care which. If this should happen to you, you can run PROBE directly. Type PROBE and press enter (or rerun from PROBE menu).

1)   ADD TRACE   This is the basic function used to display graphs. It allows us to input the variables we wish to display. Note that there is a colon and the number 1 at the end of the menu line. This means that function number 1 has already been selected. All we need do is press the return key. If you want another function, you must press the key for that function number.

3)   X AXIS   This command allows us to work with the x axis. The submenu when the x-axis function key is depressed is:

```
0) Exit 2) Linear 4) Set range 5) X variable 6) Fourier
```

Let us explore the x-axis menu and see its use.

0) EXIT   This is the escape function, and it is also capable of taking us out of the entire program. It takes only about three pressings of this function to exit the program.

2) LINEAR   This is the linear axis function, which makes the axis linear regardless of the condition of the axis prior to the time that this function is chosen. You can use this function to change a logarithmic (or exponential) axis to a linear axis.

4) SET RANGE   This function allows us to change the range of the axis from one value to another. Thus this is an expander or compressor of the axis. It is similar to the time per division control of an oscilloscope. If we had a function that had most of its detail in a small area, we could set the range to cover only this area. This would allow us to examine a small area of the curve in more detail, over the entire screen.

Expanding the axis beyond the limits set by .AC is not productive since the end of the sweep is the end of the axis as well. And we will not gain anything by expanding the axis beyond this point. This function is also of use when we use the cursor controls for measurements. You must remember that if the x axis is logarithmic, we may not have a value of zero on the axis. To change the range of the x axis, we must enter two values. The following format must be used:

$$X_{min}, \ X_{max} \ \texttt{<return>} \quad or$$
$$X_{min} \ X_{max} \ \texttt{<return>}$$

Two values are always required for setting the range of any axis. The lowest value or, if used, a negative value always comes first.

5) X VARIABLE   This function allows us to work with the x axis to change its parameter values to what we need them to be. These values can be almost any of the device voltage or current values. This function is not available for the y axis. You can also calculate values for the axis, or you can enter expressions for the x axis in this function. For instance, if our x parameter is frequency, we can change the frequency parameter to radians by typing the following equation:

```
FREQUENCY*2*3.14159
```

4) Y AXIS   This command performs nearly the same functions as the x-axis command. The main difference is that the y axis cannot have its variable changed in the same way as the x axis. There are three functions in this submenu:

```
0)Exit 1)Log 4)Set range
```

Let us explore this menu.

6) FOURIER   This function performs a Fourier decomposition of the applied voltage. Nothing more will be said about this function until we use the .FOUR function of PSpice. Press 0 or EXIT to return to the main menu.

0) EXIT FUNCTION   This is the escape function; it is also capable of taking us out of the program. It takes only about three pressings of this function to exit the program. As with the x axis, use it carefully.

1) LOG   The y axis starts as a linear axis. This function allows us to change the linear axis to a logarithmic axis.

This is useful when we are making Bode diagrams of frequency versus amplitude since it makes it easier to interpret values on the curve. As with any of the logarithmic functions, if zero is included an error will be reported and the zero will have to be removed before this function will work.

4) SET RANGE   This function operates essentially the same as the x-axis set range function.

Press 0 or EXIT to return to the main menu.

You may have more than one plot on the same graph. These plots may be entered separately or at the same time as the first graph. This is done by leaving a space or entering a comma between the graph parameters you want plotted. It is easier to enter them all at once since the graph must be redrawn each time if we use the ADD FUNCTION key for each graph we want to display. An example of this type of entry is

```
add expressions or functions
V(8) DB(V(3)/V(2)) IP(R3) VR(3)
```

Be sure to leave a space or place a comma between all the entries or an error will result. We can also use the fifth function of the menu to add graphs on their own plots with their own y axis.

5) ADD PLOT   The function of the ADD PLOT command is to allow us to add additional separate plots to the screen. When this command is selected, the plot screen is divided into two separate plots if you are using the student version. If you are using either of the other versions, the command you must use is PLOT CONTROL, and you must select from a menu. The menu is

```
0)EXIT 1)ADD PLOT 4)SUPPRESS SYMBOLS
```

When the ADD PLOT function is pressed, two graphs show on the screen and a new submenu is printed on the screen. When this function is chosen, the functions above are displayed as well as functions 2 and 3: the SELECT PLOT and REMOVE PLOT functions. The SELECT PLOT function allows us to choose which of the plots on the screen to work with. The function of REMOVE PLOT is obvious.

The ADD PLOT function can be used to make up to eight graphs on one screen, but the size of the graph's y axis diminishes with each plot added. The law of diminishing returns applies after the third graph is added. Regardless of the number of graphs, the x axis will be the same for all. It is not possible to have two different x axes on the same screen.

8) HARD COPY   This function allows us to make a hard copy of the plot shown on the screen. When this function is chosen, a submenu is shown which asks us what size the plot is to be, or to abort the plot altogether. Functions 1, 2, and 3 are the options shown. Functions 1 and 2 are obvious. Function 3 allows us to set the size to either larger than or smaller than the functions 1 and 2. The sizes must be in inches for this option.

9) SUPPRESS SYMBOLS   When more than one trace is placed on the same graph, they are identified by symbols that will be printed on each of the graphs. These symbols are an open square, a solid square, a diamond, and so on. If it is not desired to have these symbols appear on the graphs, press the 9 key to suppress the symbols. (This is most useful if you do not have a math coprocessor, since it takes some time to calculate the position of the symbols.) If you are using EGA or VGA, the graphs are printed to the screen in colors and the symbols are not used on the graphs until more than four graphs have been plotted on the same screen. The symbols can be suppressed in any case. The symbols are printed to the hard copy in all cases.

Making measurements with the cursors   This is a most useful function because it allows us to make measurements of graphic differences. If you are using student PSpice, the cursors are two arrowheads. If you are using the other versions, the cursors are two sets of crosshairs that stretch across the screen horizontally and vertically. Cursor 1 is finely spaced dots, and cursor 2 is wider-spaced dots for identification. In the latest version, the cursors work with a mouse as well as with the right–left arrow keys. Measurements are made on the graph itself, and we can then use these measurements to find slopes, -3-dB points, and so on. We can see how this function works and how the measurements are made if we look at the submenu for this function. In the student and older advanced versions of PSpice, the cursors are accessed by pressing A; in the latest version, the cursor is accessed by pressing C. The cursors also have different formats in the student and advanced versions. The menu for student PSpice is

```
C1 = XX.XXEX,  Y.YYY C2 = XX.XXEX,  Y.YYY Diff = X.XXX,  Y.YYY
0)Exit 1)Active C1 2)Active C2 3)Next Trace 4)Previous Trace :0
```

The cursor menu for the advanced versions is

```
C1  = X.XXXEX    Y.YYYEY
C2  = X.XXXEX    Y.YYYEY
DIF = X.XXXEX    Y.YYYEY
```

```
0)EXIT :0
```

From the display above we see there are two cursors that can be moved, and they display their coordinates in X,Y form. Cursor 1, C1, is always active when this function is enabled. All that is necessary to use the cursor is to use the right and left arrow keys or your mouse (left button). Cursor 2, C2, can be used by pressing the number 2 key for the student version, or holding down the shift key and using the right–left arrow keys or mouse (right button) for the advanced versions.

The cursors move between the values that were calculated for each point during the sweep or other analysis that was done on the circuit. For this reason, if you look for perfect − 3-dB points, you probably will not get them. If you have used enough points in your analysis, you will come close.

If there are two traces on the screen, we can move cursors from one trace to the other. For the student version this is done by selecting option 3, next trace. For the advanced versions this is done by holding down the control (CTRL) key and pressing the right arrow key. When the traces are put on the screen with a single command line, it is possible to have one cursor on one of the traces and one on the other. In the advanced versions, the graph each cursor is on is indicated by squares around the trace symbol. If the traces were added separately, both cursors can be on only one of the graphs. If there are two plots on the screen, we can move between the plots by selecting the plot desired to be the active plot. It is not possible to have one cursor on one graph and the other cursor on the other graph.

C1 can be used to measure a first point and C2 to measure the second point for a slope, or −3-dB points. For this function the only active keys are the right/left keys in the student version. In the advanced versions, the home/end keys are also active.

Something to be careful of is cursor movement. The first cursor movement is small, but the following ones are larger. Micro-movement of the cursor can be done by waiting a short time between pressing the arrow keys. In the student version, the cursor steps are cumulative as you hold the arrow key down. The cursor can cross the screen before you know it.

So far we have discussed only the simple functions of PROBE. Aside from the basic functions of +, −, ×, and /, PROBE can do other types of operations, a list of which is shown in Table 2-2. The units of measure are either shown alongside the functions, or are the usual values associated with the function.

Using PROBE   To show how PROBE works, we will use the *m*-derived filter we used before. We need only modify the netlist by removing the .PLOT command, replacing it with .PROBE, and changing the number of points per decade to 25. Modify

TABLE 2-2

Operation	Meaning
ABS(X)	$|X|$
SGN(X)	$+1$ (if $X > 0$), 0 (if $X = 0$), -1 (if $X < 0$)
SQRT(X)	$\sqrt{X}$
EXP(X)	e^X
LOG(X)	$\ln(X)$
LOG10(X)	$\log(X)$
DB(X)	$20 \log |X|$
PWR(X,Y)	$|X|^Y$
SIN(X)	$\sin(X)$ answer in radians
COS(X)	$\cos(X)$ answer in radians
TAN(X)	$\tan(X)$ answer in radians
ATAN(X)	Answer in radians
d(Y)	Derivative of Y with respect to x axis variable
s(X)	Integral of X over the x-axis variable
AVG(X)	Continuous average of X
RMS(X)	Continuous rms average of X

the filter netlist as shown below and rerun it. The table generated in the output file will now contain 76 frequencies and the voltage and phase data for the frequencies. The netlist for the filter with the PROBE command added is reprinted here for convenience.

```
*M-DERIVED FILTER DEMO
VIN 1 0 AC 1
R1 1 2 50
L1 2 3 19.1U
C1 2 0 .0039U
C2 2 3 .0033U
C3 3 0 .0039U
RL 3 0 50
.AC DEC 25 1E4 1E7
.PRINT AC V(3) IP(R1)
.PROBE
.END
```

The "add trace" function is selected automatically when PROBE is run. Press the return key and type in

```
VDB(3)
```

In a few seconds we should have a graph of the dB output versus frequency for the voltage at the output of the filter. Figure 2-5 shows this output graph.

When the trace is added, we get the same menu with a new function added, the number 2 in the student version and REMOVE TRACE in the advanced version. If we plot a curve that we find is not needed, we can remove it using this function. The

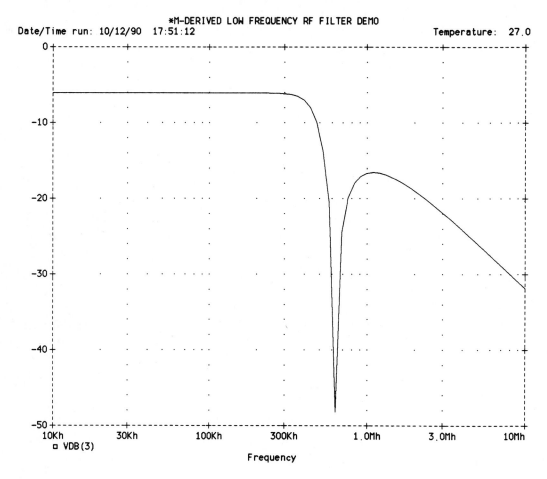

*M-DERIVED LOW FREQUENCY RF FILTER DEMO
Date/Time run: 10/12/90 17:51:12   Temperature: 27.0

Figure 2-5

submenu for this function asks which trace you want to remove or if you want to remove all the traces.

Rescaling axes.   If you have noticed, there is a long section of the curve where nothing is happening; that is, the filter has a constant output. Most of the action occurs in a narrow range of the curve toward the higher-frequency end. We can get a better idea of what is happening if we examine this section of the curve only. We can use the x-aXIS function to rescale the axis so that we can see a smaller region with better detail.

For the m-derived filter let us rescale the x axis to have only frequencies between 300 kHz and 3 MHz. These are the frequencies where the action is occurring. Press the 3 (X-AXIS) key and then the 4 (SET RANGE) key. Enter the two values as shown, then press return and the plot will be rescaled to fit the new axis values.

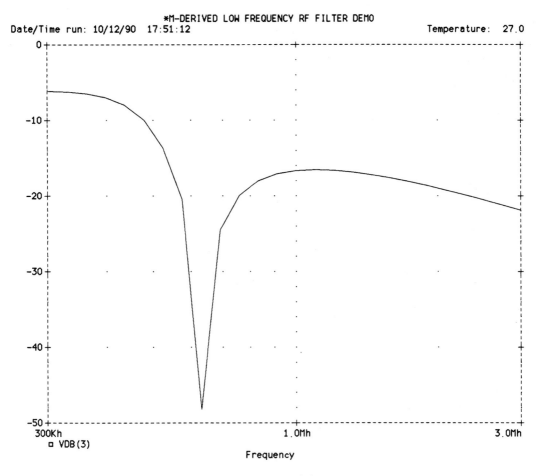

Figure 2-6

3E5,3E6 <return> or 3E5 3E6 <return>

Figure 2-6 shows the result of this change in the *x* axis. The figure shows the chart in a narrower range, but the range still includes all of the necessary data to determine any quantity we need. (An example of desired quantities is the −3-dB frequency and the frequency of maximum attenuation of the filter.) We shall not do anything to the *y* axis of the filter.

If we want to see how the phase of the output voltage changes with frequency, and we were to put this on the same plot as the decibel output curve, we would find that the phase plot occupied a much larger range than the decibel output. The phase plot would be clearly visible, but the decibel output curve would be almost a flat line. This is not good if we want to determine phase shift at a specific frequency. To get around this, we can use the ADD PLOT command of the menu. To access this command in the advanced versions, you use the PLOT CONTROL function and then press the ADD PLOT key. In the student version, simply press 5 and a new, blank plot, appears. The decibel output curve is on the bottom plot and the top plot is selected. You then use the ADD GRAPH function as you usually would. Press the 1 OK TRACE key and type in

IP(RL)

This graph will now show the phase shift of the filter versus frequency on only this graph. This will allow you to make useful comparisons of events at different frequencies.

Figure 2–7 shows the addition of the phase-angle graph to the output voltage graph. We will use the cursors to measure the -3-dB point and the frequency of maxi-

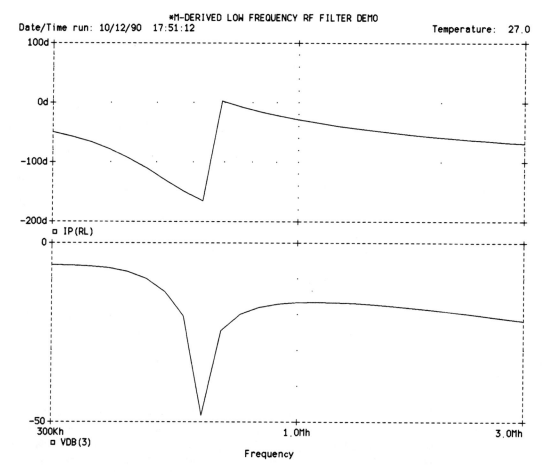

Figure 2-7

mum attenuation of the *m*-derived filter. If you are using the student version of PSpice, press A. If you are using the advanced version, press C. Then using the right/left arrow keys, or mouse, move cursor one to the frequency of maximum attenuation. For the student version, press 2 to make the second cursor active. In the advanced version hold down the shift key, or use the mouse and right button, and use the arrow keys for the second cursor. Move the second cursor to the -3-dB point for the filter. For the student version the menu portion of your screen should look like this:

```
C1=631.0E3,-41.18 C2=458.7E3,-9.015 Diff=172.2E3 39.16
0) Exit 1) Active C1 2) Active C2 3) Next Trace 4) Previous
Trace :0
```

If you are using the more advanced versions, the menu portion will look like this:

C1	= 631.1E3	−41.18
C2	= 458.7E3	−9.015
DIF	= 172.2E3	39.16

EXIT 0

The difference shown represents the subtracted values of the two measurements. This concludes our tour of the basic functions of PROBE.

2.11. Z_{in} AND Z_{out} OF NETWORKS

The input and output impedance of any network are important for us to know. We may wish to achieve maximum power transfer and match the signal source impedance but not overload the signal source.

2.11.1. DC Input and Output Resistance, .TF

The dc input and output impedance of a network are found by using the command .TF. This is the transfer function command; it evaluates the circuit for the small-signal gain and input and output impedance. The format for the .TF command is

```
.TF <output variable> <input source>
```

The output variable may be any output of the network being analyzed. The input source must be an independent voltage or current source. If the input source is a current source, PSpice will not analyze an open circuit, which the current source is. A large resistance must be placed across the source. We can analyze the input and output impedance across R_L and V_1 of Example 3 of Chapter 1 to see how .TF works. The netlist is

```
*TWO VOLTAGE SOURCE PROGRAM.
V1  1  0  10
R1  1  2  20
RL  2  0  100
V2  3  2  10
R2  3  0  10
.TF V(2) V1
.END
```

When this netlist is run, the output should be as shown below.

```
V(2)/V1                       = 3.125E - 01
INPUT RESISTANCE AT V1        = 2.909E + 01
OUTPUT RESISTANCE AT V(2)     = 6.250E + 00
```

The gain of the circuit is less than 1, which is normal for lossy resistive networks. The input resistance is taken with the voltage source disconnected and looking at the input of the system, including V_2. (Remember that V_2 is a short circuit; its resistance is zero ohms.) The output impedance is taken looking across the load resistor with the input voltage source still in the circuit.

2.11.2. AC Input Impedance

The input impedance of ac networks is not quite as simple as the dc network impedances since the input and output impedance vary with frequency. The .TF function does not work for ac circuits; therefore, we must use other techniques. PSpice calculates the magnitude and real and imaginary components of the ac voltages and currents, and we can use these values to find the impedances of the circuit under analysis.

We shall continue using the series m-derived filter and we will determine the input and output impedance of the circuit. The filter netlist is reproduced here for convenience.

```
*M-DERIVED FILTER DEMO
VIN  1  0  AC  0
RIN  1  2  50
L1   2  3  31.8U
C1   2  0  .0039
C2   2  3  .0033U
C3   3  0  .0039U
RL   3  0  50
.AC DEC 25 1E4 1E7
.PROBE
.END
```

If you remember from basic courses in electricity, the magnitude of the input impedance of an ac circuit is found using the magnitudes of the current and the applied voltage. The input impedance of the filter is

$$|Z_{in}| = \frac{|V_{in}|}{|I_{in}|} \left\lfloor \tan^{-1}\frac{X}{R} \right.$$

We will show both the impedances and phase angle for the circuit. In all we shall find four quantities: the magnitude of the input impedance, the real part of the input impedance, the imaginary part of the input impedance, and the phase angle of the input impedance. The magnitude of current and voltage is the default value for ac voltage and current (you must specifically ask for the real and imaginary components). All that is necessary to get the magnitude of the input impedance is to use these values. Thus we need only type in

V(2)/I(RIN) or VM(2)/IM(RIN)

and the magnitude of the input impedance will be displayed as in Figure 2–8.

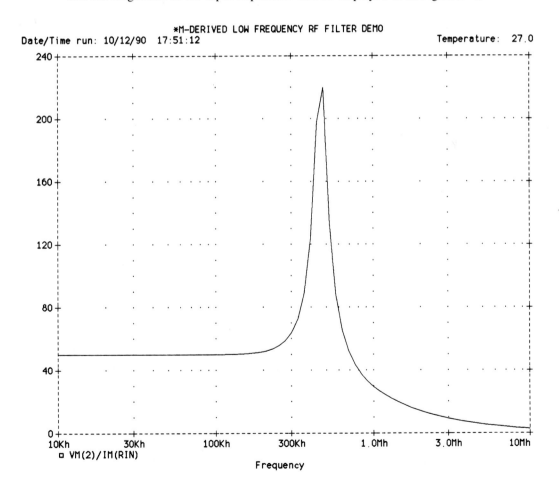

Figure 2-8

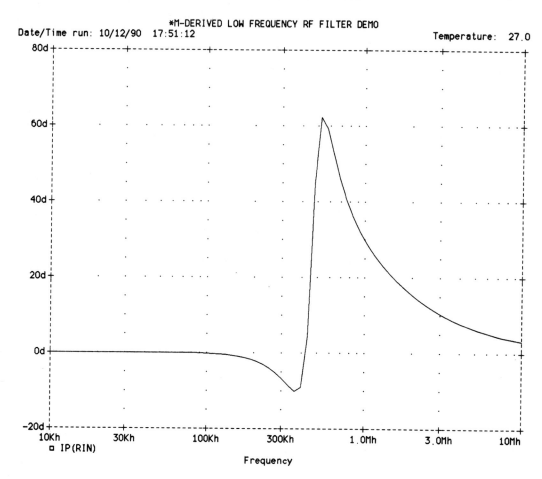

Figure 2-9

The phase angle of the input impedance is the same as the phase angle of the current in the source resistance. We can display this value by typing

IP(RIN)

and this is shown in Figure 2–9.

The real part of the input impedance of the filter will change with frequency. This is because of the resonant circuit in series between the input and output of the filter. To determine this value we cannot simply say that $R = E/I$. This is because the total input current is composed of two parts, the real component and the imaginary component. Looking at the circuit from the input, the magnitude of the current in the circuit changes depending on frequency. Since the input voltage to the filter is 1 V, we do not need to multiply by voltage for this calculation. We also must write the equation in a specific manner. We can determine the real part of the input impedance from the following calculation using PROBE:

$$\frac{IR(RIN)}{(I(RIN)*I(RIN))}$$

Note that the numerator is the real part of the current, and the denominator is the magnitude of the current squared. (See Appendix A at the back of the book for a development of how the currents involved are determined.) The parentheses are required around the entire denominator. The graph of the real part of the input impedance is shown in Figure 2–10. The imaginary part of the input impedance is found by using only the imaginary part of the current in R_1. The same type of formula is used for the

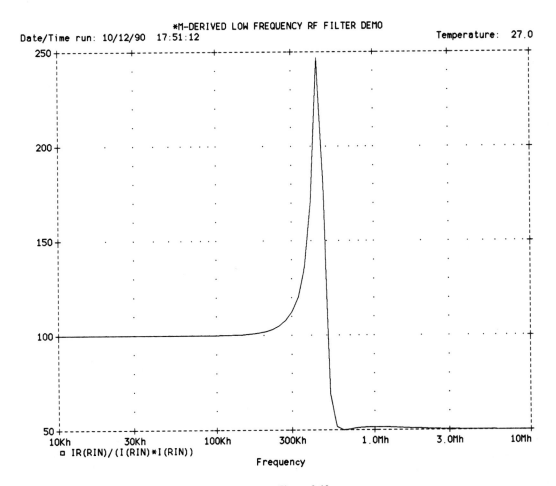

Figure 2-10

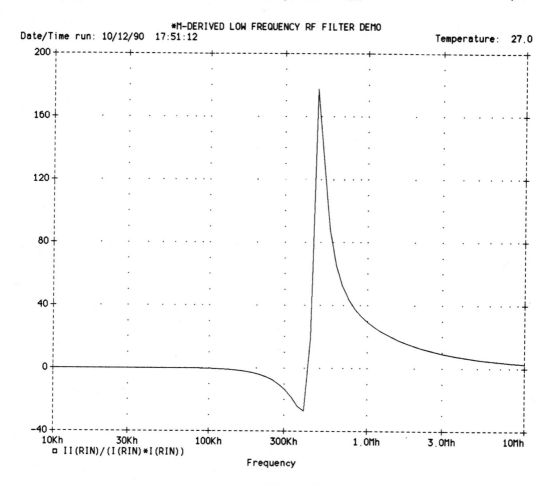

*M-DERIVED LOW FREQUENCY RF FILTER DEMO

Date/Time run: 10/12/90 17:51:12   Temperature: 27.0

□ II(RIN)/(I(RIN)*I(RIN))

Frequency

Figure 2-11

impedance, as used above. This impedance is shown in Figure 2–11 and is found from the function

$$\frac{II(RIN)}{(I(RIN)*I(RIN))}$$

2.11.3. AC Output Impedance

The ac output impedance is not as straightforward as the input impedance but works in exactly the same way. The difference comes about because we must look from the output toward the input of the circuit. When we do this, the voltage generator for the circuit is a short circuit, and the input resistance appears across the input capacitor in the circuit.

In order not to load the output of the filter, we can use a current generator across the load resistance to induce a voltage in the system. If we are smart, the current generator will produce 1 ampere, and any voltage that is generated is also the value of the impedance of the circuit. To do this calculation we will need to set the input voltage source to zero volts but not remove the generator. For these measurements we must assure that positive current flows into the output terminal of the filter. This ensures that the phase of the output current is correct. We should now change the netlist for the *m*-derived filter to reflect the devices needed for the output resistance measurements. The netlist is

```
*M-DERIVED FILTER DEMO
VIN 1 0 AC 0
Iin 0 3 AC 1
RIN 1 2 50
L1 2 3 31.8U
C1 2 0 .0039
C2 2 3 .0033U
C3 3 0 .0039U
RL 3 0 50
.AC DEC 25 1E4 1E7
.PROBE
.END
```

The magnitude of the output impedance is simply the voltage across the capacitor that is created by the current generator. To produce this impedance, we type in the following:

$$\frac{V(3)}{I(Iin)} \quad \text{or} \quad \frac{VM(3)}{IM(Iin)}$$

This impedance is shown in Figure 2–12.

The phase angle of the output voltage is simply the phase angle of the voltage at the output. To produce the phase angle curve, type

$$VP(3)$$

The phase angle graph is shown in Figure 2–13.

The real and imaginary parts of the output impedance can be shown by methods similar to the input impedance real and imaginary parts. To obtain these functions, type

$$VR(3)/(I(Iin)*I(Iin))$$

and

$$VI(3)/(I(Iin)*I(Iin))$$

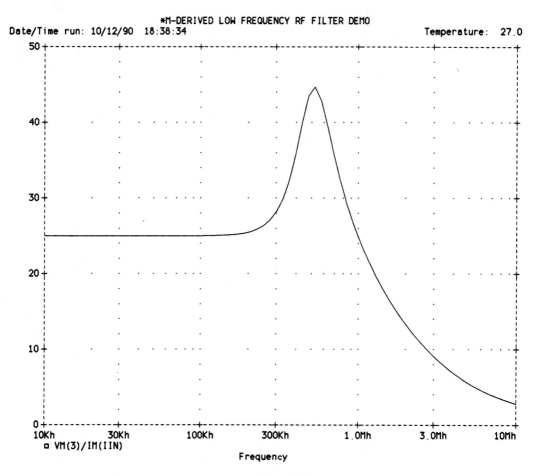

Figure 2-12

where VR and VI are the real and imaginary voltages across the output of the filter. The real and imaginary components of the output impedance are shown in Figures 2–14 and 2–15.

SUMMARY

PSpice has four types of dc sweep available that allow the changing of the dc input voltage or current to a circuit: the linear sweep; the decade sweep, which is a logarithmic type of sweep; the octave sweep, which is also logarithmic; and the list sweep, which calculates the circuit parameters at voltages or currents specified by the user.

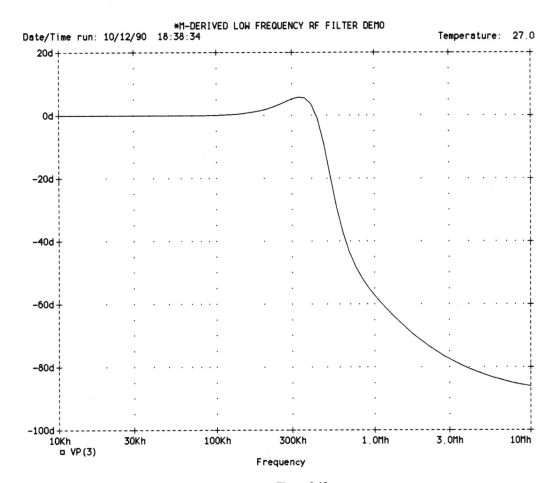

Figure 2-13

Zero cannot be used in either of the logarithmic types of sweep. The decade and octave sweeps allow for the number of points per decade or octave to be specified by the user, but PSpice selects the calculation points.

PSpice has three types of ac sweep available: the linear sweep; the decade sweep, which is logarithmic; and the octave sweep, which is also logarithmic. Zero cannot be an axis value for either the decade or the octave sweep. The decade and octave sweeps allow for the number of points per decade or octave to be specified by the user, but PSpice selects the calculation points.

PSpice can store the points calculated in the output file in the form of a table or graph; the user specifies the order in which the values are stored. The .PRINT and .PLOT statements work for both the dc and ac form of analysis.

PSpice has a graphics postprocessor called PROBE. PROBE is similar to a soft-

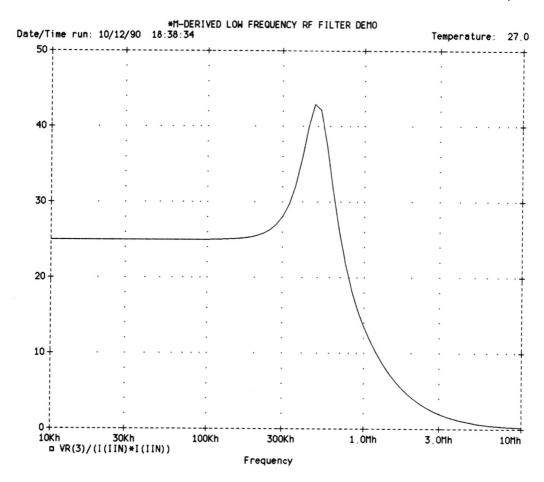

Figure 2-14

ware oscilloscope in that it can display the frequency response and dc or ac levels in a circuit. Calculations can be made using the parameters calculated by PSpice and displayed on the PROBE screen. There are a few different versions of the PROBE screen, depending on the version of PSpice used.

PROBE has a cursor function that allows the close measurement of the circuit under test output waveforms. Measurements for voltage or dB level and frequency response can be made on passive circuits using the cursor function.

The dc input and output resistance is calculated by the .TF command in PSpice. This calculation includes the small-signal dc gain of the circuit being tested.

The ac input and output impedance of passive circuits can be calculated using the magnitude and real and imaginary components of ac current and voltage calculated by PSpice.

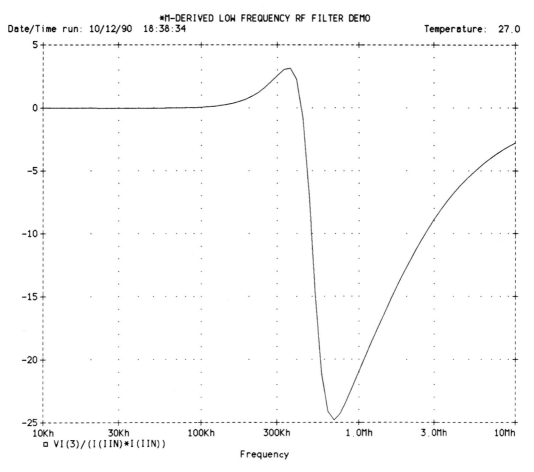

Figure 2-15

SELF-EVALUATION

For the circuits shown in Figures P2–1 to P2–5, plot the frequency response and phase angle of the input and output. Also find the input and output impedances both real and imaginary.

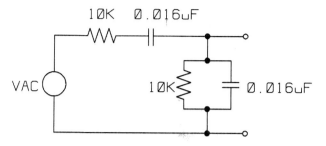

Figure P2-1

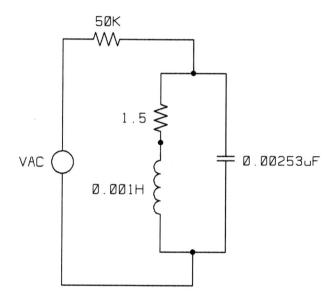

Figure P2-2

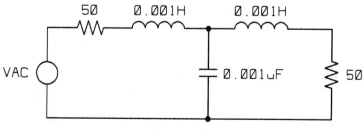

Figure P2-3

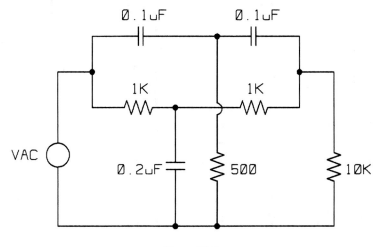

Figure P2-4

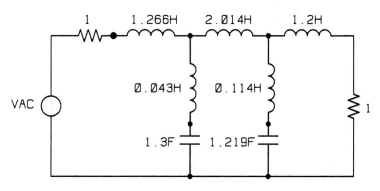

Figure P2-5

3

Device Models and Amplifiers

OBJECTIVES

1. To learn how models can be used to change parameters of passive devices.
2. To investigate what parameters of passive devices can be changed.
3. To learn how to use the dc sweep to change the model parameters.
4. To learn the use of the .TEMP command to change the temperature at which calculations are made, and to use the dc sweep to change temperature in a circuit and observe the results of the calculations for each temperature in the sweep.
5. To learn the use of models with active devices, and to learn how to change the model parameters stored in a library file.
6. To learn how to make active device models using the PSpice built-in default parameters for diodes, transistors, JFETs, and MOSFETs.
7. To apply the models and analyses learned previously to simple amplifiers to produce the frequency response and gain and phase curves of the amplifiers.

INTRODUCTION

In electronics we work with many devices. We work with amplifiers, oscillators, digital devices, op amps, and so on. In this chapter we investigate how PSpice can help us analyze circuits using models of various types of components, such as resistors, capacitors, inductors, transistors, and FETs. Models are one of the options that give PSpice its flexibility. We can change the parameters of a device, using a model, and study the effect the change has on the operation of our circuit. In some cases the models are complex and would take many lines of netlist, not to mention typing. The model for a complex component can be stored in a file called a library. To access the model, just a few lines in the netlist are needed, simplifying the netlist. There is a library file on your disk. We shall talk about library files as we need them, and we shall make new library files as we need them.

3.1. .MODEL; GENERAL REQUIREMENTS FOR MODELS

We begin by studying how to make models for passive devices, and the commands that are necessary to invoke all the models. For any type of component model, the .MODEL statement is required, regardless of whether or not the component model is in the netlist itself or in a library. This statement indicates to PSpice that the parameters of a device will be set by some model, and all the devices that reference this model will be treated in the same way. To use the models there are some simple rules to be followed:

1. More than one device of the same type may reference a given model in a netlist using the .MODEL statement, but only one type of device can be referenced by any .MODEL statement.
2. More than one .MODEL statement can be used, and each .MODEL statement must have a different model name. An example is having several .MODEL statements for resistors. As many resistors as necessary can reference model number 1 and all resistors that reference this model statement will be treated the same. As many as necessary may reference model number 2, and all of these are treated the same; and so on. Only one group of parameter changes is allowed per statement; that is, no .MODEL statement can have more than one group of parameters to be changed. The names of resistor models in each .MODEL statement must be different.
3. No device may reference a model statement that is not for that type of model. A resistor cannot reference an inductor, capacitor, or other type of model.
4. The .MODEL statement must include all the parameters you wish to change during a sweep of the device. It is not mandatory to include all the parameters of the device being changed. Any parameter not specified causes the program to assume a default value for that parameter.
5. No component may reference more than one model in the netlist. That is, the model for a resistor may reference one and only one of the models for resistance, regardless of how many resistor models are called out in the netlist. For this reason parameters to be changed must be specified completely.

3.2. SPECIFYING MODELS

The type names associated with passive component models are:

1. Resistor   RES
2. Capacitor   CAP
3. Inductor   IND

In all model statements for passive components one of these names must be used. The general form of the statement that invokes a passive model is

```
.MODEL <model name> <type name> ([<parameter value>· · ·])
```

The first name in the statement, ⟨model name⟩, should be a name that reflects the type of device being changed. If the device is an inductor and there are several inductors in the circuit, the name might be IND1. In any case, the name should be mnemonic. In all cases, the name must start with a letter but may be followed by numbers or numbers and letters if it is desired.

The second name in the statement, ⟨type name⟩, must be one of the three types listed: IND, CAP, or RES. The ([⟨parameter value · · ·]) must be all the parameters that are to be changed for the device. (Device parameters are specified in each of the examples that follow.) A tolerance may also be specified for the parameter. This is done by separating the tolerance by a space and specifying the tolerance.

Working circuits will not be used in the discussions of the form of the models. We use complete circuits when all of the model types have been discussed.

3.3. RESISTOR MODELS

Resistors are the simplest of the devices to model. The type of resistor that is generally used is the constant-value type although there are voltage-dependent resistors (VDRs), thermistors, and others. These, too, may be modeled using the parameters available in PSpice. We will consider only the constant-resistance type in the development of the resistor model. The resistor has the following parameters that may be changed using the .MODEL statement:

R	resistance multiplier	default = 1	
TC1	linear temperature coefficient	default = 0	$°C^{-1}$
TC2	quadratic temperature coefficient	default = 0	$°C^{-2}$
TCE	exponential temperature coefficient	default = 0	$\%/°C$

Notice that three of the available parameters are temperature coefficients. We usually think of the resistor as having no attributes other than resistance. This is not the case.

If the change of resistance with temperature is important, we must specify these coefficients.

The complete syntax for modeling a resistor is as follows:

```
R<name> <+node> <-node> [model name] <value>
```

You already understand most of this line. The only change is when a model is used. The model name must be added to the resistor statement. Notice that the value of the resistor is the last parameter to be named. This is because the resistor will now be calculated from one of the following two formulas:

$$value \cdot R \cdot [1 \; + \; TC1 \cdot (T \; - \; T_{nom}) \; + \; TC2 \cdot (T \; - \; T_{nom})^2] \tag{1}$$

$$value \cdot R \cdot 1.01^{TC}E \cdot (T - T_{nom}) \tag{2}$$

Formula (1) uses the linear and quadratic temperature coefficients of the resistor. These coefficients are specified in the .MODEL statement. Notice that if either of the temperature coefficients is not specified in the model, that part of the equation is zero. If neither of the coefficients is called out, the resistance is only the base value. Also important is the fact that the value of the resistance may be either positive or negative but may never be zero.

Formula (2) is the exponential form of the temperature coefficient. If its value is zero, the value of the resistance is equal to the base value. To specify any of the temperature values, all of the coefficients must be known for the type of resistor being used. This is usually available from the manufacturer.

An example of the use of the .MODEL statement for resistors is the following:

```
R5 6 0 RMODEL 1K
R6 7 0 RMOD 5K
.MODEL RMODEL RES (R=1)
.MODEL RMOD RES (R=2)
```

There are two different models in use here. For both of these models, all the temperature coefficients are equal to zero. Notice that the resistor lines give the value of the resistor in ohms, while the model lines give the value of the resistance as 1 and 2, respectively. It is perfectly all right to give the resistance value in either resistor lines or model lines. The values of resistance in the model statement lines are multipliers of the values in the resistor lines. The value of R6 is actually 10K because the model parameter R = 2 multiplies the 5K value by 2. All the resistors that reference the model will be multiplied by the value in the model line. As shown, the value for the resistance will be positive; the resistance can also be negative, but may never be zero. Obviously, a value of zero resistance will require infinite current, and not even PSpice can do that.

3.4. CAPACITOR MODELS

Capacitors are only a little more complex. Capacitors have one more parameter than a resistor. The capacitor has both temperature- and voltage-dependent capacitance functions. You can also specify an initial value of voltage for the capacitor. This provides an initial guess for the voltage across the capacitor in the circuit. The initial value is used in the bias calculations for the circuit.

The capacitor has the following parameters that may be changed using the .MODEL statement:

C	capacitance multiplier	default = 1	
VC1	linear voltage capacitance coefficient	default = 0	V^{-1}
VC2	quadratic voltage capacitance coefficient	default = 0	V^{-2}
TC1	linear temperature capacitance coefficient	default = 0	$°C^{-1}$
TC2	quadratic temperature capacitance coefficient	default = 0	$°C^{-2}$

If any of these parameters is not included, the parameter defaults to the value specified. The syntax for specifying a capacitor is

```
C<name> <+node> <-node> [model name] <value>
+                              [IC=<initial value>]
```

The node functions are as usual, but the model can include all of the coefficients that are listed above. There is also a new parameter that can be included, [IC = ⟨initial value⟩]. This allows us to place a voltage across the capacitor initially if we are doing an analysis that requires a preexisting condition. If a model name is included in the component line, the capacitance is calculated from the following formula;

$$\langle value\rangle \cdot C \cdot (1 + VC1 \cdot V + VC2 \cdot V^2) \cdot [1 + TC1(T - T_{nom}) + TC2 \cdot (T - T^2_{nom})] \quad (3)$$

The value shown is positive, but it may be negative. It also may never be zero. An example of two capacitor models is

```
Cout 7 0 CMOD 3.3u
C3 11 33 CMODEL 10u
.MODEL CMOD CAP(VC1=0.1 C=1)
.MODEL CMODEL CAP (C=1) IC=10V
```

Note that in the first model, the linear voltage coefficient is used. This causes the capacitance to be calculated from the formula shown above. In the second model specification, the value of the capacitance is the value specified. In addition, an initial condition of 10 V is placed across the capacitor.

3.5. INDUCTOR MODELS

The inductor is specified similarly to the capacitor and has similar coefficients that can be specified:

L	inductance multiplier	default = 1	
IL1	linear current coefficient	default = 0	ampere^{-1}
IL2	quadratic current coefficient	default = 0	ampere^{-2}
TC1	linear temperature coefficient	default = 0	°C^{-1}
TC2	quadratic temperature coefficient	default = 0	°C^{-2}

Similar to the capacitor, if any of the coefficients is not specified, the program will use the default values shown. The complete syntax for specifying an inductor is

```
L<name> <+node> <-node> [model name] <value>
+                       [IC = initial value>]
```

All of the model conditions for the capacitor also apply to the inductor. If the model name is specified, the inductance is calculated from the formula,

$$\langle value \rangle \cdot L \cdot (1 + IL1 \cdot I + IL2 \cdot I^2) \cdot [1 + TC1 \cdot (T - T_{nom}) + TC2 \cdot (T - T_{nom})^2]   (4)$$

In this case the driving function specified is a current. As before, the value specified may be either positive or negative, but never zero. An example of inductors used with models is shown below.

```
L1  9 10 LMOD 25U
L2 15 17 LMODEL 1M
.MODEL LMOD INDCL=1) IC=0.01
.MODEL LMODEL INDCL=2)
```

In the model LMOD, there is an initial condition of 10 mA applied to the inductor, and the inductance value is unchanged from the one shown. Note that in this example the value for LMODEL is L = 2. This multiplies the inductance of any inductor referencing this model by 2.

3.6. SWEEPING THE COMPONENT VALUES

At last it is time to specify how to sweep the component value. It will probably not be too big a disappointment if there is not some exotic method for specifying the component sweep. It is, in fact, the same sweep that we use for sweeping the dc voltages applied to a circuit. The same rules apply for this sweep as for any dc sweep. We specify the sweep using the .DC sweep command. The source name statement is general and can include component names as well as the source names. The general syntax for the sweep is shown below.

```
.DC <device type> <model name (R,C,L)> <start value>
+   <stop value> <increment> <nested sweep>
```

The functions R, C, and L must be in the sweep statement enclosed in parentheses and after the model name. Spaces may be left between the model name and the function, but this is not necessary. Note that a nested sweep can also be used with the component sweep.

```
R1 1 2 RMODEL 2
R2 2 0 RMODEL 1
.MODEL RMODEL RES(R = 1K)
.DC RES RMODEL(R) 1000 2000 100
```

In this example both of the resistors use the same model and are changed in 100Ω increments between 1 and 2 kΩ. Notice that R_1 and R_2 have a multiplier of 1000. It is to be understood that the multiplier has no effect during the sweep. Both resistors will be swept through the range of 1000 to 2000 Ω. An example of sweeping a resistance is shown in a netlist that sweeps a value of resistance in a simple voltage divider, changing the voltage across R_2. You should run this netlist and examine the result of changing the resistance for yourself.

```
*RESISTOR MODEL DEMO FILE
VIN 1 0 10
R1 1 2 1E3
R2 2 3 RMODEL 1
R3 3 0 1E3
.MODEL RMODEL RES(R = 100)
.DC VIN 0 10 1 RES RMODEL (R) 100 1000 300
.PROBE
.END
```

Although this is not a useful application of the model function, it shows clearly how to use the model and model sweep functions of PSpice. Note that the dc sweep is nested and will produce four curves of resistance versus voltage. One caveat that must be pointed out here is that the input voltage and current can have a value of zero. However, PSpice will abort if we try to sweep through a value of zero for the resistance.

3.7. TEMPERATURE ANALYSIS AND SWEEPING TEMPERATURE

3.7.1. Testing Specific Values of Temperature

Often, we must determine how a circuit design will perform with wide changes in the temperature of either the environment or the components. PSpice provides a method of doing this with a command statement that has the form

```
.TEMP <value> · · ·
```

where ⟨value⟩ is the temperature in degrees Celsius. The temperatures may not be specified in degrees Fahrenheit. The .TEMP statement may contain more than one temperature value. When the statement contains more than one temperature, the .TEMP statement acts as an outer loop in a nested sweep and the analysis of the system will be done at each of the temperatures. This is a discrete type of analysis, similar to the LIST type of sweep, in which only the temperatures specified will be analyzed. You have undoubtedly noticed the temperature printed on all the PROBE graphs that you have run. This is the nominal temperature that is used in PSpice for calculation of all circuits when no temperature is specified. It is 27°C. This temperature may be changed using the .OPTIONS command, but this is not usually done. The .OPTIONS command allows us to change many of the default parameter values that PSpice has built in. A list of the values that can be changed by .OPTIONS is shown in Appendix B. The default for temperature is 27°C. But it can be any value that is required for any analysis by using this command. More than one default may be set by a single command and the options can be in any order.

3.7.2. Sweeping Temperature

If it is desired to sweep the circuit temperature over a range of values, the .DC sweep function may be used. The .AC function will not sweep temperature. If there is an ac sweep in the system, a single ac sweep will be done. Then the dc sweep will be done and the temperature-sensitive component parameters will be updated and calculated at the temperatures specified. The general syntax for this type of sweep is

```
.DC TEMP <start value> <stop value> <increment>
+ <nested sweep>
```

This command will perform a temperature analysis of the circuit being analyzed at the temperatures specified in the sweep. As with all dc sweeps, the temperature sweep can be used in a nested sweep. We can sweep a component, or source, and a temperature in the same command statement. The general form for a single temperature calculation is

```
.TEMP 100
```

This command performs all circuit calculations at 100°C only. The form for a temperature sweep is

```
.DC TEMP -25 50 5
```

This command will calculate all parameters of the circuit being analyzed using −25°C as a starting point and 50°C as an ending point. The increment for the calculations is 5°C.

3.8. LIBRARY FUNCTIONS OF PSPICE

This is an extremely useful function that PSpice has available for our use. There is a library file on the disks that you have been working with, but we have not had a need to use it yet. A library file is where models of devices or circuits are kept. In general, these are devices or circuits that are used often. The type of library file that you have is determined by the version of PSpice you are using. In the student version, all the devices that come with this version are kept in a library named NOM.LIB. Depending on the version you are using, you may have one or more libraries.

Models for components of any type can be stored in library files. We can set up our own models of devices and store them in a library file that we create, or we can also access the models that are already in the library files of PSpice. When models are included in a PSpice netlist, you do not need to access the library files. If you want to use devices that are in the library files, you must use a command statement in the netlist to access the library file. The general command line for library files is

```
.LIB <library name>.LIB
```

The library file that is called must be resident on the disk with the circuit file on it. When this command is used, the library file we specified is accessed. If a specific library file is not named, the library file named NOM.LIB is the library file that is accessed. NOM.LIB is the default library file. Depending on the version of PSpice you are using, NOM.LIB will contain either the devices or the names of individual libraries. In the student version, the devices are contained in the library file. In the advanced versions, NOM.LIB directs the program to the library that has the model type specified in your netlist.

We can also set up our own library files using the file extension .LIB. We can use these files just like the other library files. When we do this we must be careful not to give our library file the same name as one of the ones that is already on the disk.

3.9. USING .OP FOR OPERATING POINT DATA

Before the discussion of active device models, we should discuss a parameter that allows us to see the device operating conditions. This command is the .OP command. In all cases, when an active device is used, the device parameters are listed in the output file. This is important, but these are not the conditions under which the device was operating when the analysis was made. To see the operating conditions of the device in the circuit, it is necessary to add a single command line with .OP. You will see this used in the analysis of the simple amplifiers in this chapter.

3.10. ACTIVE DEVICE MODELS; THE DIODE MODEL

Like all other models, the diode model, has parameters that can be changed when describing the model. The diode has 14 parameters in the student version and 20 parameters in the advanced version. All of the parameters can be changed or allowed to default. We will specify only the values that we need, and let the others default. The parameters not in the student version are marked with an asterisk at the beginning of the parameter line. The parameters of a diode that may be specified in a model are shown in Table 3–1.

The diode is specified in a circuit file in the following way:

```
D<name>  <+anode>  <-cathode>  <model name>  [area value]
```

Here we go again; another parameter has been added! This parameter is the [area value] parameter. The area value is the size of the diode. It is how many diodes are paralleled together. When an area of more than one, the default value, is used, certain parameters are increased or decreased. The parameters IS, the diode saturation current; CJO, the diode junction capacitance; and IBV, the diode reverse breakdown current, are multiplied by the area value. RS, the diode bulk resistance, is divided by the area. If you use diode parameters from a data sheet and the area value is not known (and it usually will not be), the area value is allowed to default to 1.

The ⟨+ anode⟩ is the element of the diode from which current flows to the ⟨−cath-

TABLE 3-1

	Parameter	Default	Unit
IS	Saturation current	$1E-16$	ampere
N	Emission coefficient	1	
RS	Parasitic resistance	0	ohm
CJO	Zero-bias *p-n* capacitance	0	farad
VJ	*p-n* potential	1	volt
M	*p-n* grading coefficient	0.5	
FC	Forward-bias depletion coefficient	0.5	
TT	Transit time	0	second
BV	Reverse breakdown voltage	∞	volt
IBV	Reverse breakdown current	$1E-10$	ampere
EG	Bandgap voltage barrier height	1.11	eV
XTI	IS temperature coefficient	3	
KF	Flicker noise coefficient	0	
AF	Flicker noise exponent	1	
*ISR	Recombination current	0	ampere
*NR	Emission coefficient for ISR	2	
*IKF	High-injection knee current	∞	ampere
*TIKF	IKF temperature coefficient—linear	0	$°C^{-1}$
*TRS1	RS temperature coefficient—linear	0	$°C^{-1}$
*TRS2	RS temperature coefficient—quadratic	0	$°C^{-1}$

ode⟩, of the diode. When we are using a zener diode in a simple regulator, we must be sure that the anode is connected to ground for a positive voltage supply. This means that the first connection to the diode is node zero. This makes correct the connections to the diode for positive applied voltages. All of the diode parameters can be changed by changing them in a model statement. If they are not specified either in the library or by you, they will default to the values specified in the table. Diodes that are in the library specify the necessary parameters. These are listed below for the 1N759 zener diode. Note that capacitances and switching times are not usually needed for zener diodes because they are usually used in dc circuits, so these parameters are not specified.

```
.MODEL D1N759 D(IS=0.5UA RS=15 BV=11.37 IBV=0.05UA)
```

For high-speed switching applications, the 1N914 and 1N916 diodes are included in the library. The specifications for this diode are

```
.MODEL D1N914 D(IS=100E-15 RS=15 CJO=2PF TT=12NS
              BV=100 IBV=100E-15)
```

The transit time (TT) and capacitance (CJO) of the switching diode are of importance and are included for this diode. If you do not include the values of capacitance for switching diodes, you could end up calculating into a diode that has zero switching time forcing a transition in zero time. This causes calculation and convergence problems.

Zener diode sweep example.   Let us now do a simple netlist using resistors and a zener diode to produce a circuit. We shall make a simple zener diode regulator, change the input voltage, and observe the change in output voltage. The input voltage will change from zero to a value that is twice the zener voltage. We will also plot the output using PROBE. The netlist is shown below. We will use the 1N752, 5.2-V zener diode.

```
*ZENER DIODE DEMO FILE
VIN 1 0 0VOLT
R1 1 2 100OHM
D1 0 2 D1N752
.LIB
.DC VIN 0 12 .5
.PROBE
.END
```

When this netlist is run, the diode type is addressed through the library file under the name .MODEL D1N752. It is mandatory that line 4 specify D1N752. Obviously you must know in advance the component identification numbers contained in the library file. The components in the library can be determined by simply using the command

```
TYPE NOM.LIB
```

to see the library file of interest. It may be to your benefit to use this command to print the library file on the printer.

When you have run this netlist, you can see the output curve of this circuit by typing

```
V(2)
```

We can specify the parameters for a diode by researching a data book and getting parameters from the book. We can also store any diode data we specify from the book in the library file. Let's make a netlist in which we set up our own diode parameters. We will use the 1N4742, 12-V zener diode. The parameters to be used are specified below. We will also add a sweep of the load resistance used with the regulating circuit.

```
.MODEL D1N4742 D(IS=0.05UA RS=9 BV=12 IBV=5UA)
```

These data were taken from the manufacturer's data sheet for the 1N4728 through 1N4764 series zener diode. The series resistance, RS, that is used is the zener impedance at the operating current from the data sheet. The data on most data sheets are not complete, but we can find enough information to make a simple model. The netlist for this circuit is

```
*ZENER DIODE MODEL DEMO FILE.
VIN 1 0 0
R1 1 2 10
D1 0 2 D1N4742
RL 2 0 RMOD 1
.MODEL RMOD RES(R=100)
.MODEL D1N4742 D(IS=0.05UA RS=9 BV=12 IBV=5UA)
.DC VIN 4 24 .5 RES RMOD(R) 100 200 25
.PROBE
.END
```

Note that we did not use the library files for this model. You may wish to store these data in the library files for the zener diodes. You should run this netlist and observe the output for yourself.

Finally, there is one last diode model we can make. This is the default model. In this case we simply allow all the parameters of the diode to default. We specify nothing.

```
D1 1 2 DMOD
.MODEL DMOD D              ;default model
```

Notice that all that is required for a default model is to specify D. All of the parameters will default to what is listed in the table of parameters. All of the active devices have a default model that is simply the parameter defaults listed in their respective tables.

3.11. BIPOLAR JUNCTION TRANSISTOR MODEL

The bipolar junction transistor default model that is used by PSpice is the Ebers–Moll model. In all, there are 40 parameters that can be changed or defaulted in the student version of PSpice and 50 in the advanced versions. The table of parameters is shown in Table 3-2. The parameters not used in the student version are again marked with an asterisk at the beginning of the parameter line.

The method used to specify a bipolar junction transistor in a netlist is

```
Q<name>  <collector node> <base node> <emitter node>
[substrate node] <model name> [area value]
```

The collector, base, and emitter nodes must be listed in the order shown. Once again, a new parameter is introduced. The parameter is the substrate node of the transistor. The substrate is simply the surface that the transistor is built upon. If it is not specified, it defaults to ground.

If the model name specified is a transistor in the NOM.LIB file, we can utilize this device without typing in all the parameters needed for calculation. Of course, the library that the transistor is in must be on the disk the netlist is on. As for the diode model, we can choose some specifications from a data book and store them in the library for later use.

An example of the parameters used for the 2N2222 transistor model in the library is shown below. Note that the parameters used do not represent all of the available parameters of the transistor.

```
.MODEL 2N2222 NPN(Is=3.108f Xti=3 Eg=1.11 Vaf=131.5
+           Bf=217.5 Ne=1.541 Ise=190.7f IKF=1.296
+           XTB=1.5 Br=6.18 Nc=2 Isc=0 Ikr=0
+           Rc=1 Cjc=14.57p Vjc=.75 Mjc=.3333
+           Fc=.5 Cje=26.08p Vje=.75 Mje=.3333
+           Tr=51.35n Tf=451p Itf=.1 Vtf=10 Xtf=2)
```

As you can see, many of the parameters listed are not available on most data sheets. You would have to go to the manufacturer to get some of them. Still we can make a suitable model using the parameters found on the data sheet.

Let us discuss several of the parameters used to make a model of a transistor. These parameters will allow us to make a usable model for any transistor. The parameters to be discussed are IS, VAF, BF, CJE, CJC, and TF.

IS is the *p-n* junction saturation current, if you remember from basic device phys-

TABLE 3-2

	Parameter	Default	Unit
IS	*p-n* saturation current	1E−16	ampere
BF	Ideal maximum forward beta	100	
NF	Forward current emission coefficient	1	
VAF(VA)	Forward Early voltage	infinite	volt
IKF(IK)	High-current beta rolloff	infinite	ampere
ISE(C2)	Base–emitter leakage saturation current	0	ampere
NE	Base–emitter leakage emission coefficient	1.5	ampere
BR	Ideal maximum reverse beta	1	
NR	Reverse current emission coefficient	1	
VAR(VB)	Reverse Early voltage	infinite	volt
IKR	Reverse beta high-current rolloff	infinite	ampere
ISC(C4)	Base–collector leakage saturation current	0	ampere
NC	Base–collector leakage emission coefficient	2.0	
RB	Zero-bias (maximum) base resistance	0	ohm
RBM	Minimum base resistance	RB	ohm
IRB	Current halfway to RBM	infinite	ampere
RE	Emitter ohmic resistance	0	ohm
RC	Collector ohmic resistance	0	ohm
CJE	Base–emitter zero-bias *p-n* capacitance	0	farad
VJE(PE)	Base–emitter built-in potential	0.75	volt
MJE(ME)	Base–emitter *p-n* grading factor	0.33	
CJC	Base–collector zero-bias *p-n* capacitance	0	farad
VJC(PC)	Base–collector built-in potential	0.75	volt
MJC(MC)	Base–collector *p-n* grading factor	0	
XCJC	Fraction of C_{bc} connected to R_b	1	
CJS(CCS)	collector-substrate zero-bias *p-n* capacitance	0	farad
VJS(PS)	collector-substrate built-in potential	0.75	volt
MJS(MS)	collector-substrate *p-n* grading factor	0	
FC	Forward-bias depletion capacitance coefficient	0.5	
TF	Ideal forward transit time	0	second
XTF	Transit time bias dependence coefficient	0	
VTF	Transit time dependency on V_{bc}	infinite	volt
ITF	Transit time dependency on I_c	0	ampere
PTF	Excess phase at $1/(2\pi f_t)$ Hz	0	degree
TR	Ideal reverse transit time	0	second
EG	Bandgap voltage (barrier height)	1.11	eV
XTB	Forward and reverse beta temp coefficient	0	
XTI(PT)	IS temperature effect exponent	3	
KF	Flicker noise coefficient	0	
AF	Flicker noise exponent	1	
*ISS	Substrate *p-n* saturation current	0	ampere
*NS	Substrate *p-n* emission coefficient	1	
*TRE1	RE temperature coefficient—linear	0	$°C^{-1}$
*TRE2	RE temperature coefficient—quadratic	0	$°C^{-2}$
*TRB1	RB temperature coefficient—linear	0	$°C^{-1}$
*TRB2	RB temperature coefficient—quadratic	0	$°C^{-2}$
*TRM1	RBM temperature coefficient—linear	0	$°C^{-1}$
*TRM2	RBM temperature coefficient—quadratic	0	$°C^{-2}$
*TRC1	RC temperature coefficient—linear	0	$°C^{-1}$
*TRC2	RC temperature coefficient—quadratic	0	$°C^{-2}$

ics, this is leakage current in a *p-n* junction, from which the forward current is specified by Shockley's equation:

$$I_d = I_s \left[\exp\left(\frac{V_d}{n} V_t\right) - 1 \right] \qquad (5)$$

The default value for this current is $1E-16$ A and this value sets the base–emitter voltage of the transistor. A germanium transistor would have IS in the $1E-9$ range, giving a V_{be} value of about 0.3 V. As a rule of thumb, we can say that the base–emitter junction voltage drop increases by approximately 40 mV for each magnitude power of 10 of the leakage current. Thus for silicon transistors, the leakage current should be approximately $1E-16$ A. This is the same as the default value given for transistors.

VAF is a parameter called the Early effect voltage that was first reported by J. M. Early in the 1950s. It relates to the base current curves of a transistor not being perfectly horizontal. The base current curves have a slight upward slant that increases the collector current with increasing collector–emitter voltage. Since there is more than one base current curve, the Early voltage is the point where all the curves meet on the negative collector voltage axis. Although known to be negative, the voltage is always stated as a positive voltage. The slope of the base current curves is given as $1/h_{oe}$, the output conductance of the transistor. The default for this voltage value is infinity, or curves that are horizontal. An example of the Early effect voltage is shown in Figure 3–1.

BF is the forward beta of the transistor (I_c/I_b). The default value for this parameter is 100.

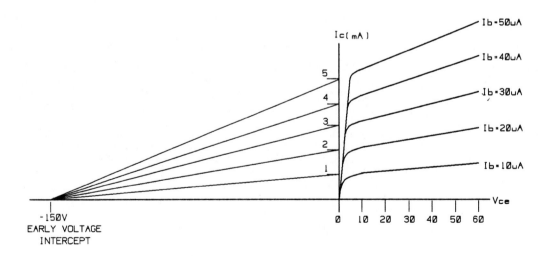

APPROXIMATE CURVES FOR A TRANSISTOR

Figure 3-1

Any model made using these parameters will be usable for low frequencies only. This is because no capacitances or rise or fall times have been included in the transistor specifications. If you need to use the model for higher frequencies, you must include the capacitances CJC and CJE. A good start for these capacitances is C_{ib} and C_{ob} of the transistor. You must alter these values until the proper f_t of the transistor is obtained. The value of f_t used can be found if the .OP command is used.

The forward transit time (TF) of the transistor is not usually given on the data sheets for transistors. It is related to the unity-gain frequency of the transistor, f_t. A reasonable approximation of this parameter can be found using this equation:

$$\mathrm{TF} = \frac{1}{2\pi f_t}$$

The rest of the parameters of the transistor can be allowed to default to the values contained in the PSpice model.

Let us make a model of the 2N3904 using data sheet values for the transistor. Then we will make a netlist to plot the collector I/V curves. We will call the model 2N39XX because the 2N3904 is in the library of PSpice already. We will also make a netlist for the 2N3904 using the library file. We will run them together and plot the collector families on the same graph and note any differences. The data sheet model for the 2N3904 is

```
.MODEL Q2N39XX NPN(BF = 300 VAF = 48.78 CJE = 8P CJC = 4P
              TF = 530.52P)
```

A brief word of explanation here about the parameters used. VAF was chosen on the basis of the slope of h_{oe} being the mean between the minimum and maximum data sheet values. BF was chosen as the mean of the minimum and maximum values of beta from the data sheets. CJE and CJC are directly from the data sheet values. TF was calculated from the gain–bandwidth product frequency. All of the other parameters, including IS, have been allowed to default.

The circuit used to produce the curves is shown in Figure 3–2. The netlist for making the curves is

```
*2N3904 V-I CURVE DEMO FILE.
Iin 0 1 0
Q1 2 1 0 Q2N3904 ;transistor model
V1 2 0 0
.MODEL Q2N3904 NPN(Is=6.734f Xti=3 Eg=1.11 Vaf=74.03
+    Bf=416.4=Ne=1.259 Ise=6.734f Ikf=66.78m Xtb=1.5
+    Br=.7371 Nc=2 Isc=0 Ikr=0 Rc=1 Cjc=3.638p Vjc=.75
+    Mcj=.3085 Fc=.5 Cje=4.493p Vje=.75 Mje=.2593
+    Tr=239.5n Tf=301.2p Itf=.4 Vtf=4 Xtf=2 Rb=10)
.DC V1 0 15 0.1 Iin 1E-5 1E-4 2E-5
.PROBE
```

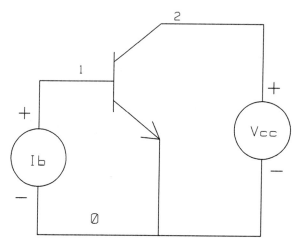

Figure 3-2

```
.END
*2N39XX V-I CURVE DEMO FILE
Iin 0 1 0
Q1 2 1 0 Q2N39XX ;transistor model
V1 2 0 0
.MODEL Q2N39XX NPN (BF=200 VAF=48.78 CJE=8P CJC=4P
+                   TF=530.52P)
.DC V1 0 15 0.1 Iin 1E-5 1E-4 2E-5
.PROBE
.END
```

This netlist will produce the collector family of V/I curves from PROBE when the value asked for is IC(Q1). To show both families on a single graph, use the number 2 function shown on the first menu for PROBE. The family of curves is shown in Figure 3–3.

A nested sweep is used in these netlists. The collector voltage is swept from 0 to 15 volts for each value of base current. When the plotted parameter is IC(Q1), the collector family of curves is displayed on the screen of the monitor. Note the slight upward tilt of the base current curves. This tilt is produced by the Early-effect voltage discussed earlier. Also note that the curves for the model we have made are not the same as the curves of the model from the library. This is due to the fact that we let all of the parameters other than those specified default.

The curves that we produced are good for design, but there are other curves and data that can be produced. Let us apply a simple calculation to the 2N3904 only. You

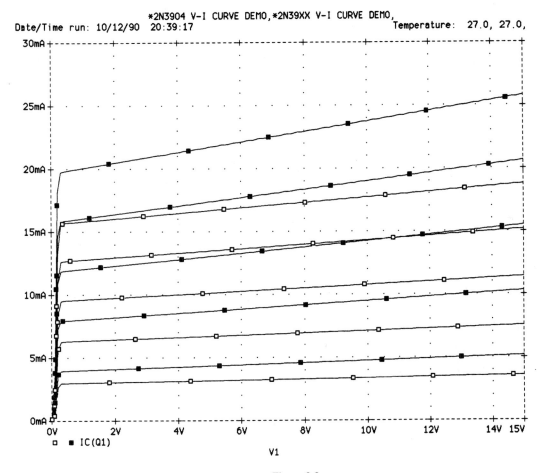

Figure 3-3

should remove the curves for the two transistors and generate a single curve for the 2N3904. We will put a load line on the plot of the characteristic curves. This is done simply by subtracting the collector–emitter voltage from the supply voltage and dividing by the desired load resistance. For this demonstration R_L is 1000 Ω.

$$(15-V(2))/1000$$

When this equation is input to PROBE, a load line should appear on the collector family of curves. We can add more load lines if we wish, to see the effect of different load resistances. The result of this is shown in Figure 3–4.

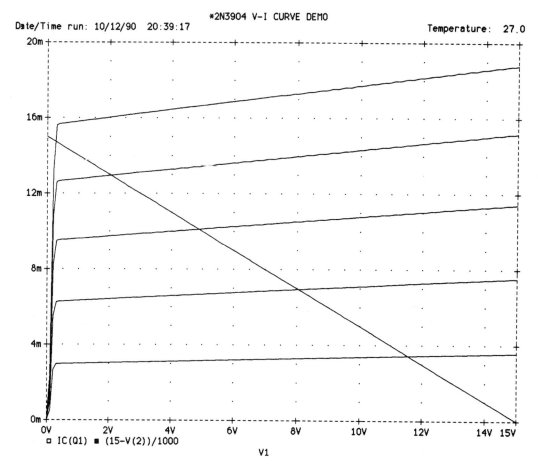

Figure 3-4

3.12. AMPLIFIER ANALYSIS USING .OP

Having the transistor curves, we could design an amplifier using them and then analyze the amplifier using PSpice. Let us, then, turn our attention to the use of PSpice to analyze a simple single-stage transistor amplifier. The amplifier will use both the 2N3904 and the 2N39XX model that we have made. We will look at the differences between the amplifiers using each of the transistor models. We will also use the command, .OP. In all cases when an active device is used, the device data are stored in the output file, but when the .OP command is used, the conditions under which the active device is operating are also stored.

   Since the procedures for transistor amplifier design are well known, we shall not go through them here. The schematic diagram for this amplifier is shown in Figure 3–5.

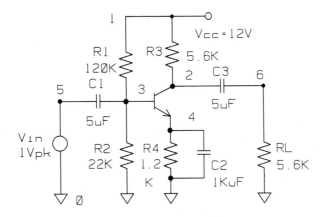

NOTE: ALL GROUND CONNECTIONS ARE NODE ZERO.

Figure 3-5

```
SINGLE STAGE AMPLIFIER 2N3904 MODEL
VCC 1 0 12
VAC 5 0 AC 1
R1 1 3 120K
R2 3 0 22K
R3 1 2 5.6K
R4 4 0 1.2K
RL 6 0 5.6K
C1 5 3 5U
C2 4 0 1000U
C3 2 6 5U
Q1 2 3 4 Q2N3904
.LIB
.AC DEC 25 1 1E9
.OP
.PROBE
.END
*SINGLE STAGE AMPLIFIER 2N39XX MODEL
VCC 1 0 12
VAC 5 0 AC 1
R1 1 3 120K
R2 3 0 22K
R3 1 2 5.6K
R4 4 0 1.2K
RL 6 0 5.6K
C1 5 3 5U
C2 4 0 1000U
C3 2 6 5U
Q1 2 3 4 Q2N39XX
```

```
.MODEL Q2N39XX NPN(BF=200 VAF=48.78 CJE=8P CJC=4P
+                     TF=530.52P)
.AC DEC 25 1 1E9
.OP
.PROBE
.END
```

The frequency response and dB output of the amplifier for both models is shown in Figure 3–6. The output curves show that amplifiers perform in substantially the same manner. The major difference in the amplifiers is in the high-frequency cutoff point. There is some difference between the two amplifiers. The 2N3904 model has the higher cutoff frequency. The difference in voltage gain of the two amplifiers is less than 1 dB.

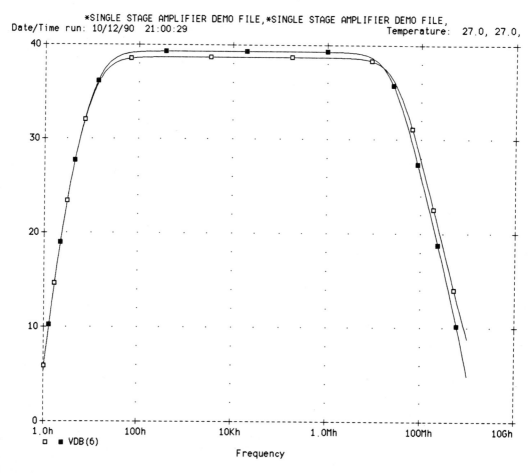

Figure 3-6

This is not significant. The models will perform in substantially the same way in any circuit.

The dc output file of this netlist also contains important data about the amplifier. All of the dc node voltages are contained in the file, as is information regarding the operating point of the transistors. The .OP command line in the netlist caused operating conditions for the transistor to be stored in the file. You can use this information to analyze the dc operating conditions for the transistor to see if the conditions produced are the ones you calculated in the design. You can also use this file to determine whether or not you need to change the capacitors specified in the library file for the transistor you made. The output file of this amplifier is shown in Figures 3–7 and 3–8.

```
******* 01/04/80 ******* Evaluation PSpice (Jan. 1988) ******* 14:22:15 *******

*SINGLE STAGE AMPLIFIER DEMO FILE

****      CIRCUIT DESCRIPTION

***********************************************************************
VCC 1 0 12
VAC 5 0 AC 1
R1 1 3 120K
R2 3 0 22K
R3 1 2 5.6K
R4 4 0 1.2K
RL 6 0 5.6K
C1 5 3 5U
C2 4 0 1000U
C3 2 6 5U
Q1 2 3 4 Q2N39XX
.MODEL Q2N39XX NPN(BF=150 VAF=48.78 CJE=8P CJC=4P TF=530.52P)
.AC DEC 25 1 1E9
.PROBE
.OP
.END
******* 01/04/80 ******* Evaluation PSpice (Jan. 1988) ******* 14:22:15 *******

*SINGLE STAGE AMPLIFIER DEMO FILE

****      BJT MODEL PARAMETERS

***********************************************************************
          Q2N39XX
          NPN
    IS    100.000000E-18
    BF    150
    NF    1
    VAF   48.78
    BR    1
    NR    1
    CJE   8.000000E-12
    CJC   4.000000E-12
    TF    530.520000E-12
```

Figure 3-7

```
******* 01/04/80 ******* Evaluation PSpice (Jan. 1988) ******* 14:22:15 *******

*SINGLE STAGE AMPLIFIER DEMO FILE

****        SMALL SIGNAL BIAS SOLUTION        TEMPERATURE =   27.000 DEG C

*******************************************************************************
   NODE    VOLTAGE      NODE    VOLTAGE      NODE    VOLTAGE      NODE    VOLTAGE

(    1)   12.0000  (     2)    7.3590  (    3)    1.7670  (     4)    1.0005

(    5)    0.0000  (     6)    0.0000


      VOLTAGE SOURCE CURRENTS
      NAME            CURRENT

      VCC          -9.140E-04
      VAC           0.000E+00

      TOTAL POWER DISSIPATION    1.10E-02   WATTS

******* 01/04/80 ******* Evaluation PSpice (Jan. 1988) ******* 14:22:15 *******

*SINGLE STAGE AMPLIFIER DEMO FILE

****        OPERATING POINT INFORMATION        TEMPERATURE =   27.000 DEG C

*******************************************************************************

**** BIPOLAR JUNCTION TRANSISTORS

NAME          Q1
MODEL         Q2N39XX
IB            4.96E-06
IC            8.29E-04
VBE           7.67E-01
VBC          -5.59E+00
VCE           6.36E+00
BETADC        1.67E+02
GM            3.20E-02
RPI           5.22E+03
RX            0.00E+00
RO            6.56E+04
CBE           3.05E-11
CBC           1.98E-12
CBX           0.00E+00
CJS           0.00E+00
BETAAC        1.67E+02
FT            1.57E+08

         JOB CONCLUDED

         TOTAL JOB TIME          9.88
```

Figure 3-7 (cont'd)

```
******* 01/04/80 ******* Evaluation PSpice (Jan. 1988) ******* 14:22:25 *******

*SINGLE STAGE AMPLIFIER DEMO FILE

****     CIRCUIT DESCRIPTION

***********************************************************************************
VCC 1 0 12
VAC 5 0 AC 1
R1 1 3 120K
R2 3 0 22K
R3 1 2 5.6K
R4 4 0 1.2K
RL 6 0 5.6K
C1 5 3 5U
C2 4 0 1000U
C3 2 6 5U
Q1 2 3 4 Q2N3904
.MODEL Q2N3904 NPN(IS=6.734F XTI=3 EG=1.11 VAF=74.03 BF=416.4 NE=1.259
+               ISE=6.734F IKF=66.78M XTB=1.5 BR=.7371 NC=2 ISC=0 IKR=0
+               RC=1 CJC=3.638P VJC=.75 MJC=.3085 FC=.5 CJE=4.493P
+               VJE=.75 MJE=.2593 TR=239.5N TF=301.2P ITF=.4 XTF=2 RB=10)

.AC DEC 25 1 1E9
.PROBE
.OP
.END
 ******* 01/04/80 ******* Evaluation PSpice (Jan. 1988) ******* 14:22:25 *******

 *SINGLE STAGE AMPLIFIER DEMO FILE

****     BJT MODEL PARAMETERS

***********************************************************************************
           Q2N3904
           NPN
     IS    6.734000E-15
     BF    416.4
     NF    1
     VAF   74.03
     IKF   .06678
     ISE   6.734000E-15
     NE    1.259
     BR    .7371
     NR    1
     RB    10
     RBM   10
     RC    1
     CJE   4.493000E-12
     MJE   .2593
     CJC   3.638000E-12
     MJC   .3085
     TF    301.200000E-12
     XTF   2
     ITF   .4
     TR    239.500000E-09
     XTB   1.5
```

Figure 3-8

******* 01/04/80 ******* Evaluation PSpice (Jan. 1988) ******* 14:22:25 *******

*SINGLE STAGE AMPLIFIER DEMO FILE

****   SMALL SIGNAL BIAS SOLUTION   TEMPERATURE =   27.000 DEG C

****

   NODE   VOLTAGE   NODE   VOLTAGE   NODE   VOLTAGE   NODE   VOLTAGE

(1)   12.0000 (2)   7.0018 (3)   1.7397 (4)   1.0787

(5)   0.0000 (6)   0.0000

        VOLTAGE SOURCE CURRENTS
        NAME   CURRENT

        VCC   -9.780E-04
        VAC   0.000E+00

        TOTAL POWER DISSIPATION   1.17E-02 WATTS

******* 01/04/80 ******* Evaluation PSpice (Jan. 1988) ******* 14:22:25 *******

*SINGLE STAGE AMPLIFIER DEMO FILE

*****   OPERATING POINT INFORMATION   TEMPERATURE =   27.000 DEG C

****
**** BIPOLAR JUNCTION TRANSISTORS

NAME   Q1
MODEL   Q2N3904
IB   6.42E-06
IC   8.93E-04
VBE   6.61E-01
VBC   -5.26E+00
VCE   5.92E+00
BETADC   1.39E+02
GM   3.41E-02
RPI   4.69E+03
RX   1.00E+01
RO   8.88E+04
CBE   1.67E-11
CBC   1.91E-12
CBX   0.00E+00
CJS   0.00E+00
BETAAC   1.60E+02
FT   2.91E+08

        JOB CONCLUDED

        TOTAL JOB TIME   12.14

Figure 3-8 (cont'd)

A further example of Pspice analysis of amplifiers is a two-stage, voltage-series feedback amplifier with poles at 20 Hz and 20,000 Hz. We will plot output curves from the amplifier and investigate the correctness of biasing for the amplifier. The specifications for the two-stage amplifier are as follows:

Bandwidth	20 to 20,000 Hz
Voltage gain	25(18 dB)
Feedback	Voltage series
Input resistance	>10 kΩ
Output load resistance	1 kΩ across 800 pF
Source resistance	10 kΩ
Transistors	2 each—2N3904
Supply voltage	24 V dc

What we notice first is that we are to use the 2N3904 transistor. This transistor is included in the library file. We first need to nodalize the circuit and assign the voltage sources their values. The circuit nodes are shown in Figure 3–9a and b.

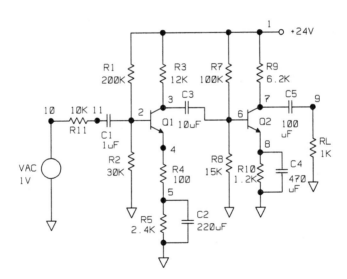

NOTE: ALL GROUNDS ARE NODE ZERO.

Figure 3-9A

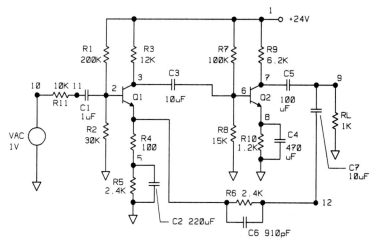

NOTE: ALL GROUNDS ARE NODE ZERO.

Figure 3-9B

The netlist for the analysis follows. Note that this netlist is done in two parts, one for the amplifier using no feedback and one including feedback.

```
*TWO-STAGE AMPLIFIER DESIGN DEMO. (NO FEEDBACK)
V1 1 0 24
V2 10 0 AC 1
R1 1 2 200K
R2 2 0 30K
Q 1 3 2 4 Q2N3904
R3 1 3 12K
R4 4 5 100
R5 5 0 2.4K
R7 1 6 100K
R8 6 0 15K
Q2 7 6 8 Q2N3904
R9 1 7 6.2K
R10 8 0 1.2K
R11 10 11 10K
RL 9 0 1K
C1 11 2 1U
C2 5 0 220U
C3 3 6 10U
C4 8 0 470U
C5 7 9 100U
```

```
.AC DEC 10 1 1E6
.PROBE
.LIB
.END
*TWO-STAGE AMPLIFIER DESIGN DEMO. (FEEDBACK)
V1 1 0 24
V2 10 0 AC 1
R1 1 2 200K
R2 2 0 30K
Q1 3 2 4 Q2N3904
R3 1 3 12K
R4 4 5 100
R5 5 0 2.4K
R6 4 12 2.4K
R7 1 6 100K
R8 6 0 15K
Q2 7 6 8 Q2N3904
R9 1 7 6.2K
R10 8 0 1.2K
R11 10 11 10K
RL 9 0 1K
C1 11 2 1U
C2 5 0 220U
C3 3 6 10U
C4 8 0 470U
C5 7 9 100U
C6 4 12 910P
C7 9 12 10U
.AC DEC 10 1 1E6
.PROBE
.OP
.LIB
.END
```

This netlist develops a large amount of data in the output file. If you are using the student version, or do not have a hard disk, you should make sure that your data disk is as clear as possible. The output file and the PROBE file are both needed to analyze this amplifier. The dc voltage values in the output file will be slightly off. This is due to the biasing resistors that were selected, which are the closest standard values to the ones calculated. Other resistor value combinations may give closer results (you can try other values). The frequency response is wider than that specified. To get the desired response, we would add some capacitance to the circuit. The frequency response curves of the amplifier are shown in Figures 3–10 and 3–11. There are many more data developed in the output file for this amplifier than are shown in the figures. You need to look at this file for yourself.

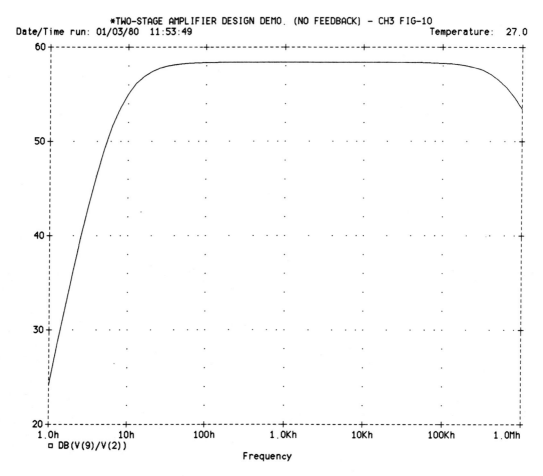

Figure 3-10

An important point to make here is that PSpice does not know that distortion will occur if the supply voltage is exceeded by the output voltage value of the amplifier. In a frequency response analysis, all devices are considered to be linear. This means that the devices can supply unlimited output voltage or current. Mostly, we want to know the output for a given input, or the gain of the amplifier. Figures 3–10 and 3–11 show the output in dB, and this value does not change, regardless of the input and output voltages.

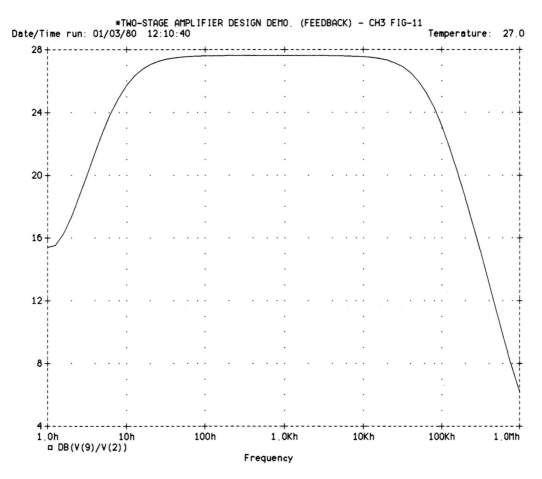

Figure 3-11

3.13. FET MODELS

The JFET is quite simple to model since it consists of an ohmic bar of silicon with a diode embedded in it. The diode is always reverse biased in order for the JFET to operate linearly. There are 14 parameters in the student version of PSpice, and 21 parameters in the advanced versions. The parameters are shown in Table 3–3. Again, the parameters not available in the student version are marked with an asterisk.

The method used to specify the JFET in a netlist is

```
J<name> <drain node> <gate node> <source node>
<model name> [area value]
```

TABLE 3-3

	Parameter	Default	Unit
VTO	Threshold voltage	−2.0	volt
BETA	Transconductance coefficient	1E−4	ampere/volt
LAMBDA	Channel-length modulation	0	volt^{-1}
RD	Drain ohmic resistance	0	ohm
RS	Source ohmic resistance	0	ohm
IS	Gate *p-n* saturation current	1E−14	ampere
PB	Gate *p-n* potential	1	volt
CGD	Gate–drain zero-bias *p-n* capacitance	0	farad
CGS	Gate–source zero-bias *p-n* capacitance coefficient	0	farad
FC	Forward-bias depletion capacitance coefficient	0.5	
VTOTC	VTO temperature coefficient	0	volt/°C
BETATCE	Beta exponential temperature coefficient	0	%/°C
KF	Flicker noise coefficient	0	
AF	Flicker noise exponent	1	
*N	Gate *p-n* emission coefficient	1	
*ISR	Gate *p-n* recombination current parameter	0	ampere
*NR	Emission coefficient for ISR	2	
*ALPHA	Ionization coefficient	0	volt^{-1}
*VK	Ionization knee voltage	0	volt
*XTI	IS temperature coefficient	3	
*M	Gate *p-n* grading coefficient	0.5	

The nodes must be in the order shown.

Of the parameters specified for the JFET, only three are needed to make a reasonable model. Of the three, one can be allowed to default if it is desired. The minimum necessary parameters are:

1. VTO   threshold voltage (pinch-off voltage)
2. BETA   transconductance coefficient
3. IS   gate *p-n* saturation current

Of these parameters, IS can be allowed to default. Of course, to make the model workable, we must have a .MODEL command for the JFET. The model can be in a library or in the netlist if desired. The syntax for the JFET model is

```
.MODEL <model name> NJF( <parameter values> · · ·)
.MODEL <model name> PJF( <parameter values> · · ·)
```

The NJF and PJF simply mean *n*- and *p*-channel devices. For the model we will construct we will use the 2N5951, for which the following parameters are available:

$$I_{gss} = 5 \text{ pA}$$
$$V_{GS(off)} = -3.5 \text{ V}$$
$$I_{DSS} = 10 \text{ mA at } V_{DS} = 10 \text{ V}$$

Two of these values can be directly entered into our model. The drain current is given by the equation $I_D = \beta(V_{GS} - V_{TO})^2$. The value of β can be approximated knowing that I_{DSS} will occur when the gate–source voltage $V_{GS} = 0$ V. The formula for this calculation is

$$\beta = \frac{I_{DSS}}{(V_P)^2} = \frac{0.01}{(3.5)^2} = 0.000816$$

Our model will then look as follows:

```
.MODEL NJFET NJF (VTO=-3.5 IS=5E-12 BETA=8.16E-4)
```

and this can be placed in a library file called JFET.LIB. Let's make a simple circuit that will produce the drain family of curves of a FET using the model we have just developed. The circuit is as shown in Figure 3–12.

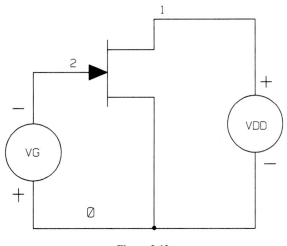

Figure 3-12

The netlist to produce the drain family of curves is

```
*JFET DRAIN CURVES DEMO FILE
VD 1 0 0
VG 2 0 0
J1 1 2 0 J2N5491
.DC VD 0 15 0.25 VG 0 -3.5 0.5
.MODEL J2N5491 NJF(VTO=3.5 IS=5E-12 BETA=8.16E-4)
.PROBE
.END
```

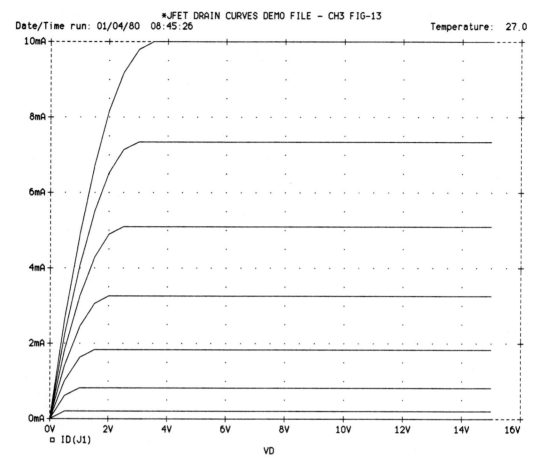

Figure 3-13

To see the drain curves in PROBE, type

$$ID(J1)$$

The drain family of curves produced by this netlist is shown in Figure 3–13.

We could also place a load line for the JFET on the drain curves using a method similar to that used for the BJT. The equation for this would be

$$I_D = \frac{V_{DD} - V_{DS}}{R_L}$$

There is another method that is used to determine the Q point of a JFET amplifier using the transconductance curve of the JFET. This curve can be generated by sweeping the gate reverse bias to produce the transconductance curve of the JFET. From this curve, the self-bias load line can be generated and shown on the curve generated by PSpice. The netlist to produce this curve is

```
*JFET SOURCE CURVES DEMO FILE
VD 1 0 15
VG 2 0 3.0
J1 1 2 0 J2N5491
.DC VG -0 -4 .1
.MODEL J2N5491 NJF(VTO=3.5 IS=5E-12 BETA=8.16E-4)
.PROBE
.END
```

When this netlist is run, we can plot the parabolic curve of the JFET by calling for the drain current

$$ID(J1)$$

The self-bias line for any source resistance can be generated by using the command

$$\frac{-V_{GS}}{R_S}$$

An example of the self-bias lines for 800-, 400-, and 200-Ω source resistances is shown in Figure 3–14.

At this point let us analyze a simple JFET audio amplifier. This amplifier uses our model. The circuit for this amplifier is shown in Figure 3–15. The netlist for this amplifier is

```
*FET AMPLIFIER DEMO FILE
VDC 1 0 15
R1 1 2 1K
R2 3 0 8.2MEG
R3 4 0 330
C1 4 0 100U
C2 2 5 10U
C3 3 6 .003U
J1 2 3 4 NJFET
VAC 6 0 AC 1
ROUT 5 0 100K
.PROBE
.AC DEC 25 1 1E9
.OP
.MODEL NJFET NJF(VTO = -3.5 IS = 5E-12 BETA = 8.1633E-4)
.END
```

The frequency response curve of this amplifier is shown in Figure 3–16.

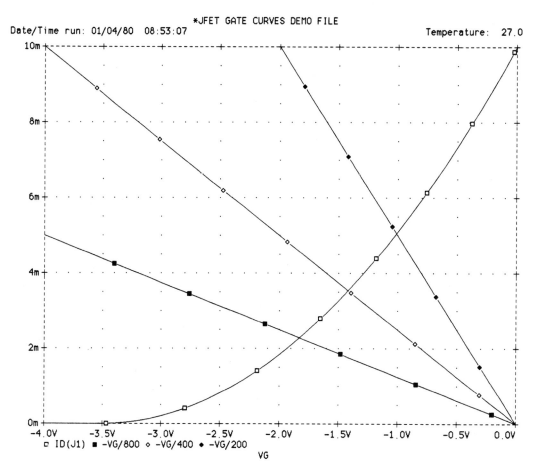

Figure 3-14

There are two more types of FET that can be modeled in PSpice: the gallium–arsenide FET and the MOSFET. The GaAs device is modeled in a manner similar to the JFET, but there are a few more parameters that can be added to the device specifications. The component line for the GaAsFET is shown below; the nodes for this device must also be in the order shown.

```
B<name> <drain> <gate> <source> <model name> [area value]
```

The MOSFET has many more parameters than those of the JFET or GaAs device. As usual, if the additional parameters are not specified, they will assume default values. The syntax for the MOSFET device line is shown below.

```
M<name> M>drain> <gate> <source> <bulk/substrate> <model name>
+        [L=] [W=] [AD=] [AS=] [PD=] [PS=]
```

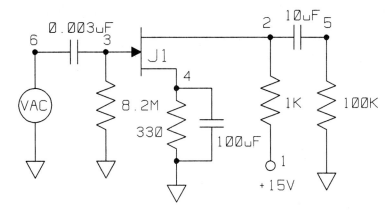

NOTE: ALL GROUND NODES ARE NODE ZERO

Figure 3-15

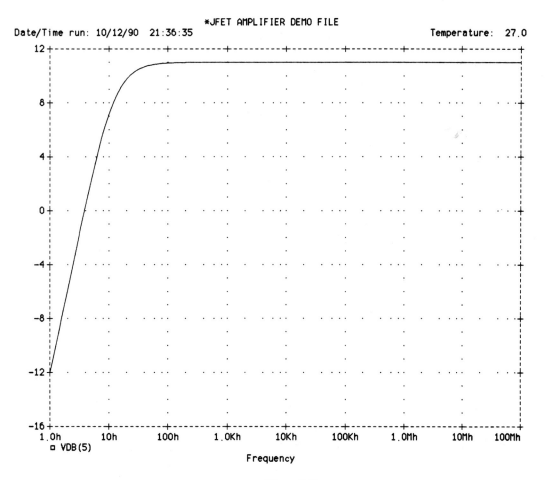

Figure 3-16

where the values L, M, AD, AS, PD, and PS are device parameters that can usually be allowed to default.

One final point to be made regarding models is that if the model line calls only for the diode (D), transistor (NPN/PNP), JFET (NJF/PJF), and so on, all of the defaults for the device will be used and we will have a generic component. An example of the model line that will cause this to occur is

```
.MODEL <model name> NPN
.MODEL <model name> NJF
```

SUMMARY

The command line starting with .MODEL is required for all components whose parameters we wish to change. There are four parameters for a resistor that can be changed. To change the parameters of a resistor, a model must be used. There are five parameters for a capacitor and inductor that can be changed. To change the parameters of a capacitor or inductor, a model must be used.

The base values of a device may be changed by the multiplier in the model statement. This is useful for scaling component values in a circuit. The temperature can be changed in two ways, either by a list form of sweep, or by using the dc sweep. The only form of dc sweep available for this function is the linear sweep. PROBE may be used to examine the results of the temperature sweep.

There are two models for the active devices. The type of model depends on the version of PSpice being used. For student PSpice, there are 14 diode parameters, 40 bipolar junction transistor parameters, and 14 JFET parameters that can be changed. For the advanced versions, there are 20 diode parameters, 50 BJT parameters, and 21 JFET parameters that can be changed.

Diodes, transistors, and JFET models can be made using certain data sheet parameters and allowing the other parameters to default. A complete default model can be made by allowing all the device parameters to default.

For BJTs, a complete family of collector curves can be generated using a simple circuit. The curves are suitable for design. The user can put load lines on the curves to determine the dc operating characteristics of an amplifier.

For JFETs, a complete family of drain curves can be generated using a simple circuit. More important, the source curve of the JFET can be generated and the necessary operating parameter lines can be drawn on this curve.

The analysis of amplifiers is easily accomplished using the ac sweep and an input voltage of 1 V peak, to determine the gain of the amplifier directly. PSpice treats all active devices as linear devices in a frequency response analysis. This allows PSpice to calculate the output without the problems of distortion that occur when we overdrive an amplifier.

The active devices used in PSpice are kept in library files. There are different library files, depending on the version being used. The student version has all the devices in a file called NOM.LIB, while advanced versions may have devices in several library files.

SELF-EVALUATION

1. The circuit of Figure P3–1 is a simple intensity modulator for a fiber optic transmission system for analog signals. You are to analyze the amplifier for the following:
   a. All dc node voltages.
   b. The frequency response and gain without the feedback capacitor (dB gain).
   c. The frequency response and gain with the feedback capacitor (dB gain).
   d. The frequency response change when the capacitor is changed from 500 to 1000 pF in 100-pF increments.

2. The circuit of Figure P3–2 is a discrete version of a type of fiber optic receiver called an integrated detector-preamp. You are to analyze this receiver for the following parameters:
   a. All dc node voltages.
   b. The frequency response of the system.
   c. The gain of the each amplifier in the circuit. The overall gain of the system must be at least 1200 for reliable detection of the optic signal.

3. A transistor has a dc β of 150, a CJC of 1 pF, and a CJE of 4 pF. The Early effect voltage can be found from the fact that $h_{oe} = 25$ microsiemens (μS) and $I_c = 0.01$ mA at zero volts. Run the collector family of curves for the transistor for $V_{cc} = 0$ to 15 V and $I_b = 10$

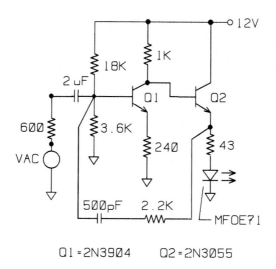

Q1 = 2N3904   Q2 = 2N3055

Figure P3-1

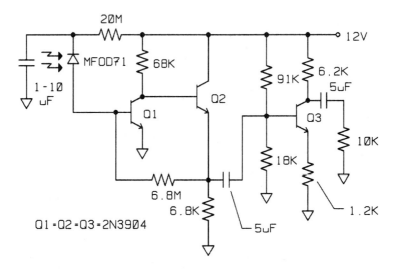

Figure P3-2

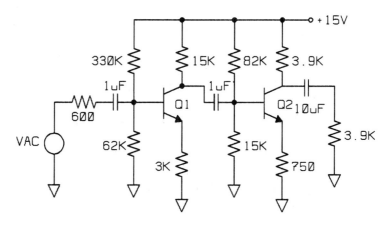

Figure P3-3

to 50 μA. Assure that the design shown in Figure P3–3 will produce the following parameter specifications:

> voltage gain = 20 dB
> frequency response = 20 Hz to 100 kHz
> V_{cc} = 15 V, I_{c1} = 0.5 mA, I_{c2} = 2 mA

Run frequency and phase response curves for this amplifier. Then make the frequency response correspond to the desired values.

4. The circuit shown in Figure P3–4 is a JFET audio voltage amplifier with a gain of 10. The JFET used has the following parameters: VTO = −4 V, BETA = 1E−4, CGS = 2 pF,

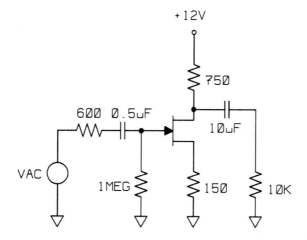

Figure P3-4

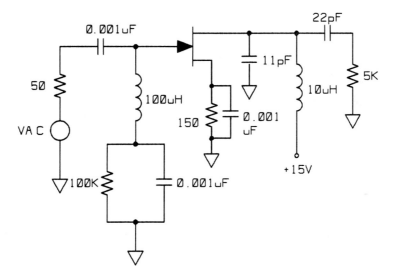

Figure P3-5

CGD $=$ 1 pF. Run the frequency and phase response curves of the amplifier and tailor the frequency response to 20 to 20,000 Hz. V_{dd} is to be 12 V.

5. The circuit of Figure P3–5 uses a 2N3819 JFET in a 15-MHz amplifier. The bandwidth of the amplifier is 1 MHz loaded. The amplifier is to have a gain of 10 dB minimum and is to have a load resistance of 50 Ω. Evaluate this amplifier for gain and bandwidth.

4

Linear and Polynomial Controlled Sources and Subcircuits

OBJECTIVES

1. To learn the use of the four types of controlled source available in PSpice.
2. To apply these sources to simple circuits and observe that these devices have similarities to amplifiers with which we are already familiar.
3. To learn to use the controlled devices in circuits that have a controlling equation. These are called polynomial controlled devices.
4. To learn to use more complex forms of controlled devices called multiple-input polynomial controlled devices.
5. To learn to use the controlled devices in circuits that produce mathematical functions.
6. To learn the principles of subcircuits as small segments of netlists that may be used several times in any netlist.
7. To make ideal operational amplifiers using the concept of subcircuits.

INTRODUCTION

So far we have used only independent sources whose output is not dependent on outside influences. PSpice is also able to work with sources that are not fixed in value—sources whose output depends on an input from some other part of the circuit. We call such sources dependent sources. For flexibility, dependent sources should also be capable of having outputs that are not linear functions of one or more inputs. In this chapter we develop four types of dependent source and some nonlinear dependent sources using these devices. We also develop some ideal operational amplifier circuits. We can place these circuits in a library file for use as we need them.

4.1. LINEAR CONTROLLED SOURCES

In analog design several types of device are used. Among these devices are four linear controlled sources. In most textbooks, these four sources are called dependent sources. Their output is a function of an input that is located elsewhere in the circuit. The sources and their names are derived from the two types of input and output possible in any circuit, either voltage or current. From these inputs and outputs, four types of device are possible. The four devices are

1. Voltage-controlled voltage source (VCVS)
2. Current-controlled current source (CCCS)
3. Voltage-controlled current source (VCCS)
4. Current-controlled voltage source (CCVS)

PSpice treats these sources as devices and gives them simple device names: E, F, G, H, respectively. To use any of these devices you must specify the device by the appropriate letter. The device letter may be followed by a name. An example of the four devices is

Edev	VCVS device
Fout	CCCS device
Gcon	VCCS device
Hamp	CCVS device

Since the devices are functions of a variable, either voltage or current, we can equate the sources in the following way:

VCVS: output voltage a function of input voltage $= \mathrm{E(Vin)}$
CCCS: output current a function of input current $= \mathrm{F(Iin)}$
VCCS: output current a function of input voltage $= \mathrm{G(Vin)}$
CCVS: output voltage a function of input current $= \mathrm{H(Iin)}$

The dependent source is usually some form of amplifying device, such as the bipolar junction transistor, which is a CCCS. Since these devices are amplifiers, it is understood that they are active devices. They are capable of delivering power to an external system or load. The devices we have been using thus far have been passive; they can only absorb power from the system. In electric circuits these dependent devices are

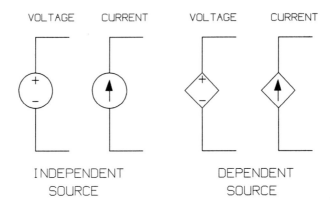

Figure 4-1

given symbols different from those of the independent source. Figure 4-1 shows the two symbols that are used in this book.

Specifying the sources is not difficult to understand, but there are some conventions that must be followed. Each of the different types of source is described briefly, followed by an example of specifying, and using, the source. The general syntax for these devices is

```
(E,F,G,H)<name> <output> <input> <controlled parameter>
```

For all the devices to be examined, the output will be nodes and the output nodes are always listed first. The inputs may be either nodes or a voltage source through which a current flows. Components such as resistors or capacitors may not be specified for the source to be measured. Depending on which type of source is being used, the controlled parameter will be either a voltage or current gain, or a resistance or conductance.

4.2. VCVS SOURCES

The VCVS device is also known as a voltage amplifier. It is similar to the noninverting op amp. It receives an input voltage; multiplies this input by some linear constant, K, the gain of the device; and then delivers an output voltage to the load connected to it. We thus have a voltage-in, amplified voltage-out device. Note that the gain of the device is a constant. This allows the input to determine what the output of the device will be.

All the sources to be discussed generally have no physical connection between their input and the source, although this is possible. This brings us to the point where we need to explain how to describe this source to PSpice. The description of the VCVS device is

$$\underset{\text{name}}{\text{Edev}} \quad \underset{\text{output node}}{\underline{5 \ 0}} \quad \underset{\text{input node}}{\underline{4 \ 0}} \quad \underset{\text{gain}}{\underline{10}}$$

Analysis of this description shows that the name of the device is Edev. Its output is between node 5 and node 0, and its input is between node 4 and node 0. The gain from input to output of the device is 10. This means that any voltage that is applied to nodes 4 and 0 is multiplied by 10 and appears across nodes 5 and 0. An equivalent mathematical expression for this device would be

$$\underset{\text{output}}{\frac{V_5}{}} = \underset{\text{gain}}{\frac{10}{}} \ \underset{\text{input}}{\frac{V_4}{}}$$

For this type of device, only one set of input nodes and one set of output nodes are allowed.

There are some simple methods that allow us to make reading the device specifications a little easier. We can use some commas and parentheses that PSpice will ignore but, cosmetically, will make the netlist easier to read. An example of this for the source just described is

```
Edev (5,0)(4,0) 10
```

As can easily be seen, the readability of the line is better than before due to the nodes being grouped within parentheses. You must remember that for these sources the output node is named first, the input node next, and the gain last. All of these quantities are separated by a space (or other device, such as a comma). The way that PSpice knows that a new parameter is being encountered is by the use of the space or the comma to separate the field values.

As an example of how the VCVS source can be used, we will analyze a circuit that is very common in industry, the equal-component Sallen–Key active filter. The filter will be of second order. The gain of the circuit will be larger than that specified to produce the maximally flat response curve. The reason for the gain being larger is to show that the VCVS will act the same as any other amplifying device in this application. The characteristic humped response will be obvious. The netlist for this filter is

```
*VCVS SALLEN-KEY FILTER DEMO
VIN 1 0 AC 1
R1 1 2 1E4
R2 2 3 1E4
R3 3 0 1E9
R4 4 0 1E9
EFLT 4 0 (3,0) 2.2    ;VCVS DEVICE
C1 2 4 0.016U
C2 3 0 0.016U
.PROBE
.AC DEC 100 1 2E4
.END
```

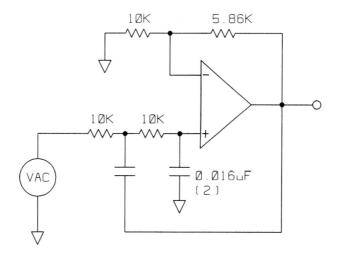

Figure 4-2

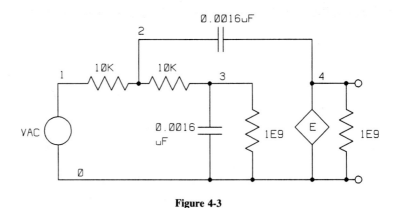

Figure 4-3

Note the use of the semicolon to indicate that a comment follows. You should run the filter netlist with different gains, including the maximally flat gain of 1.586, to see the effect of the gain on the response of the filter. The cutoff frequency for this filter is approximately 1000 Hz. The circuit for this device is shown in Figure 4-2 and its equivalent using the E device is shown in Figure 4-3. The output of the filter is shown in Figure 4-4.

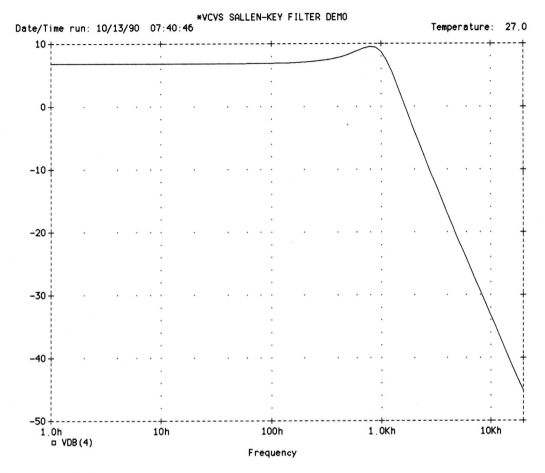

Figure 4-4

4.3. CCCS SOURCES

The current-controlled current source is similar to the bipolar junction transistor. A description of how to label a CCCS device is

$$\underset{\text{name}}{\underline{\text{Fout}}} \quad \underset{\text{output node}}{\underline{2\ 0}} \quad \underset{\text{measured device}}{\underline{\text{V1}}} \quad \underset{\text{gain}}{\underline{5}}$$

In this case the measured device is the input voltage source to the circuit. It is the current in the voltage source that is the measured parameter. The output of the CCCS is taken at nodes 2 and 0, and the gain from input to output is 5. You must understand what happens in this circuit. Following the rules for the current source, we place a large resistance, say 1 GΩ, across the current source as a path to ground. This is all right, but suppose that the current through V1 is 0.05 A. The current source will force 0.25 A through the 1-GΩ resistance. This creates the small problem of having to find a device that can produce the output voltage for this current—in this case, 250,000,000 V! When using this device, you must pay attention to the impedance levels for the circuit.

An example of the equation type from which the line shown above could be made is

$$I_5 = 5I_{V1}$$

An example netlist using the CCCS device is shown below. This is a simple resistive circuit that can be found in any text.

```
CCCS SOURCE DEMO
V1 1 0 5VOLTS
R1 1 0 1E3OHMS
F1 2 0 V1 5              ;CCCS SOURCE
R2 2 0 1E2OHMS
.DC V1 10 10 1
.PRINT DC V(1) I(R1) V(2) I(R2)
.END
```

The output file is shown in Analysis 4-1. Note that there is no physical connection between the input voltage source and the dependent source. Also note that the output of this source is 25 mA and will produce 2.5 V across the output resistor R2.

```
******* 10/08/90 ******* Evaluation PSpice (Jan. 1988) ******* 15:39:21 *******

*CCCS SOURCE DEMO - ANALYSIS 4-1

****      CIRCUIT DESCRIPTION

***************************************************************************


V1 1 0 5VOLTS
R1 1 0 1E3OHMS
F1 2 0 V1 5
R2 2 0 1E2OHMS
.DC V1 5 5 1
.PRINT DC V(1) I(R1) V(2) I(R2)
.END
```

Analysis 4-1

```
******* 10/08/90 ******* Evaluation PSpice (Jan. 1988) ******* 15:39:21 *******

*CCCS SOURCE DEMO - ANALYSIS 4-1

****      DC TRANSFER CURVES                    TEMPERATURE =   27.000 DEG C

***********************************************************************************


  V1           V(1)        I(R1)       V(2)        I(R2)

  5.000E+00    5.000E+00   5.000E-03   2.500E+00   2.500E-02


         JOB CONCLUDED

         TOTAL JOB TIME               .55
```

Analysis 4-1 (cont'd)

4.4. VCCS DEVICES

The VCCS source is similar to the field-effect transistor. In this case the output is a current and the input is a voltage. Since we have an output current that is controlled by an input voltage, we see that the units of the output are amperes per volt. This is a conductance. Since the conductance is transferred across the input to the output of the device, we call this a transconductance amplifier. A description of a VCCS device is

$$\underset{\text{name}}{\underline{\text{Gout}}} \quad \underset{\text{output name}}{\underline{\text{4 5}}} \quad \underset{\text{input node}}{\underline{(2,3)}} \quad \underset{\text{conductance}}{\underline{0.01}}$$

Analysis of this line shows that the name of the device is Gout. The output of the device is taken between nodes 4 and 5. The input of the device is between nodes 2 and 3. The conductance through which the current is measured is 0.01 siemens (S). The measured input quantity for this device is voltage, so the nodes across which the voltage is developed are specified. It is mandatory that the output conductance be given in Siemens.

An example of the equation from which this specification can be derived is

$$I_4 - I_5 = \frac{V_2 - V_3}{100} = (V_2 - V_3) \times 0.01$$

Let us examine a simple circuit that is found in any text. The circuit is shown in Figure 4-5. This is a simple voltage-controlled circuit in which the current developed by the G device is the voltage across R2 divided by 2 A. It is desired to find the voltage across R5 and the current in R3 and R5.

The only question here might be: How do we determine the conductance to use for the G device? Simple, the conductance is the reciprocal of the divisor of V1 for the

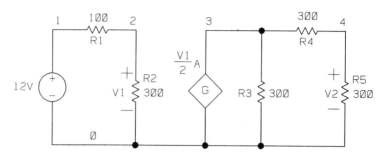

Figure 4-5

G device. This comes about by this reasoning: The voltage across R2 is 9 V. For the G device to produce 4.5 A, it is necessary that the G device multiply the magnitude of the voltage by 0.5. Thus the conductance is 0.5 of what is applied, or 4.5. Now that we know the method, it is a simple thing to predict that the voltage at node 3 will be 900 V, and the voltage at node 4 will be 450 V, the current through R5 will be 1.5 A.

```
*VCCS SOURCE DEMO FILE - ANALYSIS 4-2
VIN 1 0 12VOLTS
R1 1 2 100
R2 2 0 300
G1 0 3 2 0 .5
R3 3 0 300
R4 3 4 300
R5 4 0 300
.DC VIN 12 12 1
.PRINT DC V(2) V(4) I(R5)
.END
```

```
******* 10/08/90 ******* Evaluation PSpice (Jan. 1988) ******* 15:39:22 *******


*VCCS SOURCE DEMO FILE - ANALYSIS 4-2


****     CIRCUIT DESCRIPTION


*************************************************************************

VIN 1 0 12VOLTS
R1 1 2 100
R2 2 0 300
G1 0 3 2 0 .5
R3 3 0 300
R4 3 4 300
R5 4 0 300
.DC VIN 12 12 1
.PRINT DC V(2) I(R2) V(4) I(R5)
.END
```

Analysis 4-2

```
******* 10/08/90 ******* Evaluation PSpice (Jan. 1988) ******* 15:39:22 *******

*VCCS SOURCE DEMO FILE - ANALYSIS 4-2

****     DC TRANSFER CURVES              TEMPERATURE =   27.000 DEG C


*************************************************************************


    VIN          V(2)        I(R2)        V(4)        I(R5)

   1.200E+01   9.000E+00   3.000E-02   4.500E+02   1.500E+00



       JOB CONCLUDED

       TOTAL JOB TIME            .55
```

Analysis 4-2 (cont'd)

The output file is shown in Analysis 4-2. Obviously, we were correct in our predictions for both the voltage across, and the current through, the desired resistor. It is also obvious by the application of Kirchhoff's current law that the current in the 300-Ω resistor is the remainder of the current, or 3 A.

4.5. CCVS DEVICES

The current-controlled voltage source is a source whose output voltage depends on the input current. In this case the output is given in volts per ampere, and is therefore a resistance. Since the resistance is transferred across the input to the output, we call this a transresistance amplifier. A description of a CCVS device is

$$\underline{\text{Hout}} \quad \underline{\text{3 5}} \quad \underline{\text{V(5)}} \quad \underline{\text{20}}$$
$$\text{name} \quad \text{output node} \quad \text{device} \quad \text{resistance}$$

Analysis of this line shows that the name of the device is Hout. The output is across nodes 3 and 5, and the device whose current is being measured is V(5). The resistance through which the current flows is 20 Ω. The equation from which this type of transresistance is determined is

$$V_3 - V_5 = 20I_{V5}$$

An example of the CCVS is the simple netlist that converts input current to output voltage:

```
*CCVS SOURCE DEMO FILE
VIN 1 0 10VOLTS
```

```
RIN 1 0 1KOHM
HTEST 2 0 VIN 1KOHM        ;-CCVS DEVICE
.DC VIN 10 10 1
.PRINT DC V(2,0) I(R1)
.END
```

The output file of this circuit is shown in Analysis 4-3.

When PROBE is run for this demonstration, notice that the output is negative. This is due to the current convention used by PSpice. To make the output positive only

```
******* 10/08/90 ******* Evaluation PSpice (Jan. 1988) ******* 15:39:22 *******


*CCVS SOURCE DEMO FILE - ANALYSIS 4-3


****     CIRCUIT DESCRIPTION


*************************************************************************


VIN 1 0 10VOLTS
RIN 1 0 1KOHM
HTEST 2 0 VIN 1KOHM
R1 2 0 1KOHM
.DC VIN 10 10 1
.PRINT DC V(2,0) I(R1)
.END



******* 10/08/90 ******* Evaluation PSpice (Jan. 1988) ******* 15:39:22 *******


*CCVS SOURCE DEMO FILE - ANALYSIS 4-3


****     DC TRANSFER CURVES          TEMPERATURE =   27.000 DEG C


*************************************************************************


  VIN         V(2,0)      I(R1)

 1.000E+01  -1.000E+01  -1.000E-02



        JOB CONCLUDED

        TOTAL JOB TIME             .50
```

Analysis 4-3

requires that we change the polarity of the input voltage source, or the CCCS device. Also note that this circuit produces 1 V per milliampere. It is then a current-to-voltage converter. This type of circuit is often used in instrumentation systems to convert a transmitted telemetry current from a 5- or 20-mA current loop to a voltage.

4.6. POLYNOMIAL CONTROLLED SOURCES

The controlled sources of the preceding section can only perform the simple function of amplification. They would be more useful if we could do other operations with them. In this section we work with sources that can have more than one input and which have a defining equation. The polynomial controlled source can be used in several ways. It can have a single input or it can have several inputs. It will always have a single output. We can use it to solve equations using voltage or current as the independent variable. It can also be used to create complex output signals.

For this type of source, as in the linear controlled source, first to be described is the output nodes of the source, second is the order or dimension of the polynomial. Then the measured inputs are described, and last, the coefficients of the polynomial are described. The general syntax for the polynomial source is

```
(E,F,G,H<name>) <output nodes> POLY(X) <input nodes>
+                                       <coefficients>
```

The devices E, F, G, and H are the same as the devices described in the preceding section. They are dependent sources. The POLY(X) specification is the dimension of the polynomial that is to be used. The inputs are described next and we must have as many input pairs as there are dimensions to the polynomial. If the dimension is POLY(4), there must be four sets of input nodes or devices. The numeric coefficients of the polynomial are last to be specified. The general form of the polynomial is shown below. The polynomial coefficients must be in the order shown. Note that the polynomial always starts with a constant term and the other terms are given in ascending order of the powers of x:

$$a + bx + cx^2 + dx^3 + ex^4 + \cdots$$

To use the polynomial form, all coefficients of x up to the last nonzero coefficient must be specified. This means that any x that is not specified in the formula has a coefficient of zero and must be shown as such. If this is not done, the order of the expression will change and an incorrect solution will result. An example of how to express an equation with powers of x to PSpice is

$$1 - 3x^2 + 4x^3 + 5x^6$$

$$1 + 0x - 3x^2 + 4x^3 + 0x^4 + 0x^5 + 5x^6$$

$$\underbrace{1}_{a_0} \ \underbrace{0}_{a_1} \ \underbrace{-3}_{a_2} \ \underbrace{4}_{a_3} \ \underbrace{0}_{a_4} \ \underbrace{0}_{a_5} \ \underbrace{5}_{a_6}$$

As you can see, all powers of x, including those for whom the value of the coefficient of x is zero, have been accounted for. The controlling inputs of the polynomial are either a pair of nodes for voltage inputs, or voltage source names for current inputs. There must be as many input specifications as there are dimensions in the specification POLY(X).

The way to learn how the polynomial functions work is to do one. Let us solve a simple equation using the VCVS block. The function to be solved for is

$$y = 2x^3 - 4x$$

This is a third-order equation with the constant term and x^2 coefficient equal to zero. For this equation, the coefficients must be specified in ascending order of powers of x, and are

$$0 - 4\ 0\ 2$$

We shall sweep the polynomial from -2 to 2 V in 0.1-V increments. The netlist to implement this function is shown below. Note how the VCVS device is shown in the netlist.

```
*POLY(1) SWEEP PROGRAM FOR 2X^3 - 4X
VIN 1 0 0VOLTS
RIN 1 0 1E6OHMS
Epoly 2 0 POLY(1) (1,0) 0 -4 0 2   ;Polynomial device.
R1 2 0 1E6
.DC VIN -2 2 0.1
.PROBE
.END
```

When this netlist is run, the output at node 2 will be the curve of a third-order equation. For this equation there will be zero crossings and the curve will start at -8 V and end at $+8$ V. That these values are correct can be calculated from the equation supplied. Note that the voltage supply VIN was set to zero volts. This is acceptable, but not necessary, since the sweep will override the input value for the duration of the sweep. If we analyze this netlist line by line, we see that the input voltage source and resistance are necessary since we cannot allow the input source to float. Line 2 completes the dc path to the reference node. The polynomial device has its output at nodes 2 and 0. There is a resistor attached to node 2 to complete its dc path to the reference node. The sweep is established in line 6 and PROBE is called in line 7. Once PROBE has been executed, we can see the voltage of interest, which is the voltage at node 2.

Another operation that can be done with PROBE that has not been used yet is that of the derivative. In PROBE, we can take the derivative of a function over the range of the x axis. If we want the derivative of a function, we can do this simply by typing in a command to do the derivative. The command always starts with D.

$$D(V(2))$$

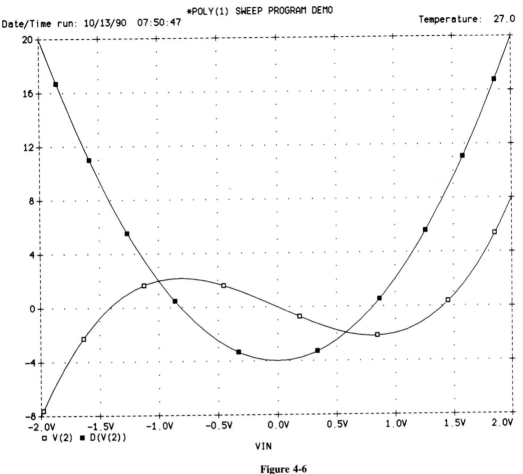

Figure 4-6

To take higher derivatives, you add more D's. You must follow each D with a paren-thesis (be careful about having the right number of closing parentheses).

$$D(D(V(2)))$$

The output of this netlist and the derivative is shown in Figure 4-6.

4.7. MULTIPLE-INPUT CONTROLLED SOURCES

The multiple-input device uses the same dependent sources that we have described be-fore. The difference is that we have more than one input to the device. This allows more interesting options.

Operations that we cannot do with a single-input device   A summing amplifier is a multiple-input device that adds all of its inputs together to arrive at an output value. PSpice has the ability to do this also. Not only can we add or subtract inputs but we can multiply and raise to a power. We can take the root of a number as well. This is done by supplying the proper feedback to the source being used. Also, we must change the way the determining function is input to the system. The multiple-input device is complex and increases in complexity as the number of inputs is increased. As an example of the type of polynomial that is input to PSpice, let us examine a function of the three inputs. This polynomial has the form

$$a_1 + a_1v_1 + a_2v_2 + a_3v_3 + a_4v_1^2 + a_5v_1v_2 + a_6v_1v_3 + \cdots$$

and this equation iterates until we have all of the voltages to the third power and all of the cross products. An example of the three input device is

$$\underline{\text{Eout}}\quad \underline{4\ 0}\quad \underline{\text{POLY(3)}}\quad \underline{(1,0)(2,0)(3,0)}\quad \underline{0\ 1\ 1\ 1}$$
$$\text{name}\qquad \text{out}\qquad \text{degree}\qquad\qquad \text{input nodes}\qquad \text{polynomial}$$
$$\text{coefficients}$$

This equation allows us to sum three inputs to arrive at an output value. The input coefficients are the last three terms of the polynomial and are equal to 1. The first coefficient is related to input (1,0), the second to input (2,0), and so on. Notice that the constant term is included and is zero for this function. The coefficients of the inputs do not have to be 1; they can have any other weights that are needed.

An example of a two-input device that would allow us to multiply two inputs is

$$\text{Eout}\ \ 3\ 0\ \ \text{POLY(2)}\ \ (1,0)(2,0)\ \ 0\ 0\ 0\ 0\ 1$$

Refer to the general equation above. Notice that for a two-input device, the first three coefficients are the constant term and the first and second input voltages. The third

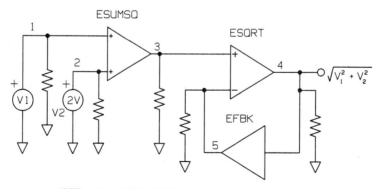

NOTE: ALL RESISTORS 1E9 OHMS

Figure 4-7

coefficient would be the second-order term of the first input voltage, that is, v_1^2. All these coefficients must be zero, accounting for the four zeros in the coefficient array. The next term is the $v_1 v_2$ term that multiplies the two voltages so that its coefficient is 1. This type of device could be used to produce amplitude-modulated waveforms if two sine waves of different frequencies were applied to the device. We shall use it for that purpose in a later chapter.

To see the multiple-input polynomial device work, we can make a simple netlist to solve the Pythagorean theorem. The netlist requires three polynomial sources—two of which are POLY(2) devices, in a feedback arrangement. The netlist file is

```
*PYTHAGORAS' THEOREM DEMO FILE
V1  1  0  0
V2  2  0  2
```

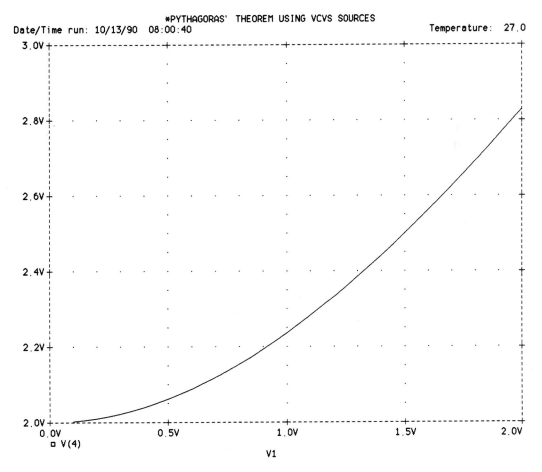

Figure 4-8

```
ESUMSQ 3 0 POLY(2) (1,0)(2,0) 0 0 0 1 0 1
ESQRT 4 0 POLY(2) (3,0)(5,0) 0 1E5 -1E5
EFBK 5 0 POLY(1) (4,0) 0 0 1
R1 1 0 1E9
R2 2 0 1E9
R3 3 0 1E9
R4 4 0 1E9
R5 5 0 1E9
.DC V1 0.1 2 .05
.PROBE
.END
```

The diagram is shown in Figure 4-7 and the output of the circuit in Figure 4-8.

4.8. SUBCIRCUIT FUNCTIONS

The subcircuit function is similar to a subroutine in programming. It is a small section of netlist that can be used several times within another netlist without rewriting it each time. If there are small circuits that you use consistently, it is at least time consuming to keyboard input a section of circuit into a design several times. It may also not be possible to input the same circuit several times either. This is due to the unique nodes used for the circuit and the unique names of components that are used.

The subcircuit is treated as a device in PSpice. As such, it is given a name—the name that is given to the subcircuit is X ⟨name⟩. This name must be used when using a subcircuit in a netlist. The subcircuit to be used in a netlist is called using the command syntax

$$X\langle name\rangle \quad \underbrace{\langle node\ 1\rangle \cdots \langle node\ N\rangle}_{nodes} \quad \langle subcircuit\ definition\ name\rangle$$

An example of a subcircuit call in a netlist is

$$XAMP \quad \underbrace{2\ 3}_{inputs}\ \underbrace{1\ 4}_{power}\ \underbrace{5}_{output} \quad UA741$$

This is a call for an op amp. In this line, XAMP is the name given to an amplifier in the netlist. The X in the netlist indicates that either there is a subcircuit in the netlist or the program will have a library file command in the netlist. The nodes given are listed in the order that the subcircuit would have in the library file. The name of the subcircuit definition is UA741.

The nodes that are specified in this command are of paramount importance. The reason for this is that these nodes are connected to the input and output nodes that are listed in the subcircuit. The subcircuit nodes and the nodes in the call for the subcircuit are usually not the same. If we do not list the nodes correctly, the program will give us the wrong answers or an error message. Node 1 is one of the nodes and node N is the last node of the system. If node 1 needs to go to the primary input of the system, the node specified must be this input point; similarly, for the output node. An important point here is that the nodes need to be in the specific order of the subcircuit. In the example above, the noninverting and inverting inputs are nodes 2 and 3, the nodes 1 and 4 are the power supplies, and node 5 is the output. If you look at the UA741 subcircuit in the library, you will find that this is the same order as the subcircuit.

The subcircuit function allows us to copy a small segment of a circuit into a netlist as many times as we like. The subcircuit may be placed anywhere in the netlist. The definition of a subcircuit to be used in a circuit has the syntax

```
.SUBCKT <definition name> <node 1> <node 2> · · ·
component lines
.ENDS
```

The ⟨definition name⟩ of the subcircuit must be the same as the name that you store the circuit under in the library. The nodes listed under .SUBCKT are the nodes that are used to connect the subcircuit to the netlist. Any node numbers that are listed in the subcircuit itself are separate from any node numbers in the netlist. The only exception to this rule is that the zero (0) node is global. It cannot be anything other than the ground reference node. Use of this function is shown in the next section.

4.9. SIMULATING IDEAL OP AMPS USING CONTROLLED SOURCES

It would be beneficial if we could make, and store in a library, some operational amplifier circuits to use as we need them in various circuits. These would be ideal amplifiers, since they would not have all the parameters of the actual amplifiers. We can do this using controlled devices to emulate the parameters of the op amp we wish to use.

There are two types of ideal op amp that can be constructed using controlled sources:

1. A simple ideal op amp with no frequency limitations.
2. A more complex ideal op amp with a single-pole rolloff. This device more closely approximates the real op amp frequency response.

The simple op amp uses only a single VCVS source. The input resistance is the only parameter that can be simulated easily with this device. An example of this type of op amp is one for the LF-155. The input resistance of this op amp is 10^{12} Ω.

```
*SIMPLE IDEAL LF-155
.SUBCKT LF_155 1 2 3
RIN 1 2 1E12
EAMP 3 0 POLY(2) (1,0)(2,0) 0 2E5-2E5
.ENDS
```

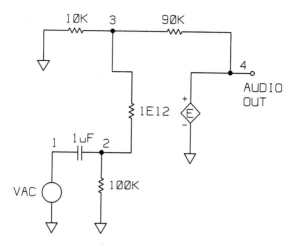

Figure 4-9

As you can see, there is not much to this ideal op amp. It should be used only for low-frequency and dc applications, since there is no upper frequency limit to this amplifier, and it is not a good predictor of the frequency response of the LF-155. The unity-gain frequency of the LF-155 is about 2.5 MHz.

We can make a simple circuit to demonstrate the use of this amplifier. The circuit is a low-frequency, noninverting audio amplifier. The low-frequency cut-in is 20 Hz, and the amplifier has a gain of 10. We will sweep from 1 Hz to 10 MHz to show that the amplifier does not limit the high-frequency response as the actual amplifier would. The amplifier is shown in Figure 4-9. The netlist of the amplifier is

```
*SIMPLE IDEAL LF-155 AUDIO AMPLIFIER
VAC 1 0 AC 1
CIN 1 2 1U
RIN 2 0 100K
XAMP 2 3 4 LF_155
RFB 4 3 90K
RF2 3 0 10K
.AC DEC 10 1 1E7
.PROBE
*SIMPLE IDEAL LF-155
.SUBCKT LF_155 2 3 4
RIN 1 2 1E12
EAMP 4 0 POLY(2) (2,0)(3,0) 0 2E5 -2E5
.ENDS
.END
```

The output of the analysis is shown in Figure 4-10. Amplifier theory says that an amplifier keeps the difference between its two inputs nearly zero. Thus, when the output

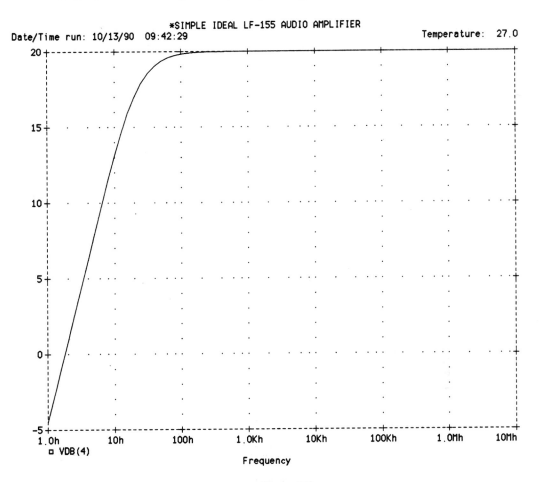

Figure 4-10

of this ideal op amp is 10 V, the feedback voltage at the other input will be 1 V and amplification ceases.

To construct the more complex ideal op amps there are certain parameters that you will need to know for the amplifier to be simulated. These parameters are:

1. The open-loop gain
2. The gain–bandwidth product
3. The open-loop cutoff frequency
4. The input resistance
5. The output resistance

The parameters should be available from data books for op amps. As a further example, let us make a more reasonable model for the LF-155. This amplifier is not in the

NOM.LIB file. So we can make a library of our own to use for the various subcircuits we make. We will call this library SUB.LIB. The parameters for this op amp are as follows;

$$A_{OL} = 200 \text{ V/mV} \qquad \text{the open-loop gain of the op amp}$$
$$\text{GBP} = 2.5 \text{ MHz} \qquad \text{the gain–bandwidth product of the op amp}$$
$$f_{OL} = 12.5 \text{ Hz} \qquad \text{the open-loop frequency response}$$
$$R_{in} = 10^{12} \text{ } \Omega \qquad \text{the input resistance}$$
$$R_{out} = 25 \text{ } \Omega \qquad \text{an approximate output resistance}$$

You must remember that the op amp that will be described using this technique is an ideal op amp. Parameters such as the slew rate for the amplifier will be missing and any design using this model will not show these effects.

To make the model of the op amp, we can use two VCVS sources. One is the input source, containing the open-loop frequency determining elements and the overall gain of the op amp; and one is the output source having a gain of 1 and the output resistance. The schematic diagram of this op amp is shown in Figure 4-11. The choice of R_{OL} and C_{OL} is arbitrary. Any combination of values that produce a pole at the 12.5-Hz open-loop frequency can be used. For this frequency response we can use the exact value of capacitance calculated, we do not have to pick a standard value. Let's use 5 KΩ for R_{OL}. The proper value of C_{OL} to produce 12.5 Hz rolloff of open-loop frequency at this value of R is 2.55 μF.

Since the gain–bandwidth product is usually the only thing that is given for the op amp, it is necessary to produce the capacitance by using the following formulas:

$$f_{OL} = \frac{\text{GBP}}{A_{OL}} = \frac{2.5 \times 10^6}{2 \times 10^5} = 12.5 \text{ Hz}$$

$$C_{OL} = \frac{1}{2\pi f_{OL} R_{OL}} = \frac{1}{2\pi(1.5)(5 \times 10^3)} = 2.55 \text{ } \mu\text{F}$$

We now have everything that is needed to make the model of the op amp. We can keep this model in library file mentioned above.

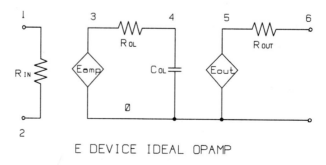

E DEVICE IDEAL OPAMP

Figure 4-11

```
*IDEAL OPAMP - LF155 ;this is a comment line, not a
.SUBCKT LF155 1 2 6  ;title line. It is not necessary
RIN 1 2 1E12         ;for the subcircuit definition.
Eamp 3 0 POLY(2) (1,0)(2,0) 0 2E5 -2E5
ROL 3 4 5K
COL 4 0 2.55U
Eout 5 0 (4,0) 1
Rout 5 6 25
.ENDS
```

Let us examine this subcircuit in some detail.

1. Line 2 identifies this as a subcircuit. The nodes of the subcircuit that connect to the outside world are shown in this line.
2. Nodes 1 and 2 are the input nodes. The input resistance, Rin, appears between these two nodes. The output is node 6. The gain and frequency determining network values are shown in lines 4, 5, and 6. The gain for this source is 200,000. The cutoff frequency elements for this source are shown on lines 5 and 6. These elements produce the 12.5-Hz rolloff of frequency for the open-loop gain.
3. The output source is on line 7. This source has a gain of 1. It provides isolation for the frequency-determining elements of the gain source.
4. The output resistance for the op amp is shown on line 8. Note that this resistance is in series with the source.

This model will emulate the low- and high-frequency operation of the op amp. It will not include the effects of slew rate, nor will this amplifier function as an oscillator. To demonstrate the use of the subcircuit, we will make a state-variable filter for 1000 Hz with a Q value of 0.707 and a Butterworth response. The circuit for this filter is shown in Figure 4-12. The netlist for the filter is

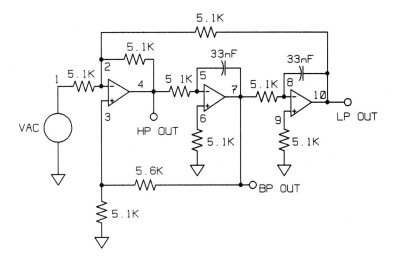

Figure 4-12

```
*LF155 STATE VARIABLE FILTER DEMO
VAC 1 0 AC 1
R1 1 2 5.1K
R2 2 4 5.1K
R3 3 0 5.1K
R4 3 7 5.6K
R5 4 5 5.1K
R6 7 8 5.1K
R7 2 10 5.1K
R8 6 0 5.1K
R9 9 0 5.1K
C1 5 7 33NF
C2 8 10 33NF
XFILT1 3 2 4 LF155
```

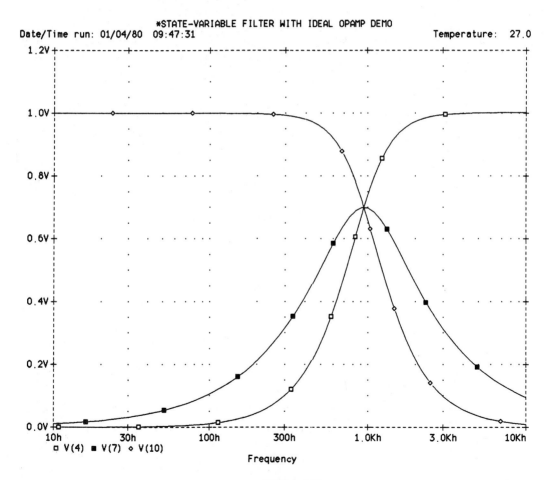

Figure 4-13

```
XFILT2 6 5 7 LF155
XFILT3 9 8 10 LF155
.AC DEC 25 10 10K
.PROBE
.LIB CIRCUIT.LIB
.END
```

The outputs of this filter for the highpass, lowpass, and bandpass functions are shown in Figure 4-13.

SUMMARY

There are four types of linear controlled devices available in PSpice: voltage-controlled voltage sources, current-controlled current sources, voltage-controlled current sources, and current-controlled voltage sources. The controlled sources are similar to voltage amplifiers, current amplifiers, transconductance amplifiers, and transresistance amplifiers.

The controlled sources can be expanded to include devices that have a controlling function, and can be used in circuits that produce many mathematical and other functions. The controlled sources can be used to construct ideal operational amplifiers. These amplifiers mimic the actual amplifier in all respects other than parameters such as the slew rate.

PSpice has a function called subcircuits. These are segments of netlists that can be called into another netlist as often as needed. Subcircuits can have only .MODEL statements as command lines.

The ideal operational amplifiers made using controlled sources can be stored in a library file as subcircuits.

SELF-EVALUATION

1. Using a F device, make a netlist that solves $y = 2x^2 - 3x - 2$. Plot the curves of the function and first derivative. The sweep is between -2 and 2 V. (*Hint:* Let the input resistor be 1 Ω.)

2. Using two E devices, make a circuit that shows the cube root of numbers between -8 and $+8$. [*Hint:* Use the POLY(2) and POLY(1) devices.]

3. The SE5512 op amp has the parameters shown below. Make an ideal op amp subcircuit for this amplifier and store it in the library.

$$\text{Rin} = 100\text{M}\Omega$$
$$\text{GBP} = 3 \text{ MHz}$$
$$\text{Av} = 200 \text{ V/mV}$$
$$\text{Rout} = 75 \text{ }\Omega$$

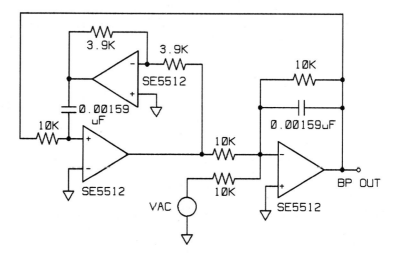

Figure P4-4

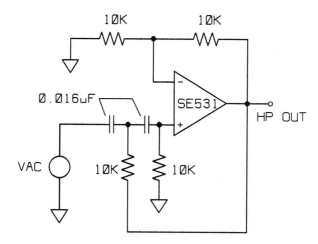

Figure P4-5

4. Using the ideal op amp of Problem 3, design and analyze the filter shown in Figure P4-4.

5. The SE531 op amp has the parameters shown below. Make an ideal op amp subcircuit for this op amp. Analyze Figure P4-5 using this amplifier.

$$Rin = 100 \text{ M}\Omega$$
$$GBP = 3 \text{ MHz}$$
$$Av = 200 \text{ V/mV}$$
$$Rout = 75 \text{ }\Omega$$

5

Transient Analysis and Signal Types

OBJECTIVES

1. To learn the use of the time analysis functions of PSpice for transient analysis.
2. To learn the required parameters for the transient form of analysis.
3. To learn the parameters of the five signal types available for transient analysis.
4. To learn the use of the coupling coefficient command to couple inductors together to form transformers.
5. To learn the methods for making nonlinear core models for transformers.

INTRODUCTION

Until now we have worked only with dc and with ac that was assumed to be a sine wave. These independent sources served the purpose at the time. However, many circuits use, or generate, signals that are not simple sine waves. This type of signal can be anything from a simple 50% duty cycle square wave to more complex pulses, staircase and modulated waves, and transient signals of various types. When these signal voltages or currents are needed we must use the transient analysis function of PSpice. The purpose of this chapter is to explain the transient analysis capabilities of PSpice and the types of signals available in PSpice for transient analysis. These signal sources are also independent sources.

5.1. .TRAN FOR TRANSIENT ANALYSIS

Transient analysis in PSpice is the method by which we simulate the operating of a circuit over some time period. Also, transient analysis is used for circuits that generate their own signals (oscillators). It is a powerful analysis tool, but like any other tool, it has pitfalls for the unwary user. We first discuss the requirements for the use of .TRAN, what it does, and some pitfalls that we may encounter.

Transient analysis is a time-domain analysis. In PSpice, time always begins at zero and moves forward. If it is required that an analysis begin at a time other than zero, a delay can be specified for the various types of signals. The syntax for the transient analysis function in PSpice is

```
.TRAN[/OP] <print time steps> <final time value> [<no print
time>]
+        [<maximum step time>] [UIC]
```

There are four parameters that can be specified for transient analysis. Of the four, two must be specified and two are optional. The ⟨print time steps⟩ and ⟨final time value⟩ are mandatory for all analyses. The [⟨no print time⟩] and the [⟨maximum step time⟩] are optional. The [/OP] command causes the transient bias point calculated for the transient analysis to be printed to the dc output file. The [UIC] command allows us to set some, or all, of the nodes to specific levels before transient analysis starts.

The ⟨print time steps⟩ parameter sets the time steps that are saved for the .PRINT and .PLOT outputs of the analysis. These are generally not the same time steps as those used for calculation in the transient analysis. When there is little or no change in the waveform being analyzed, the time steps for calculation of the output values are larger than when the calculation is being done at a corner of the waveform. For this reason the time step window will change to smaller values of time when the step is too large. This may occur several times, until a time step is found that allows calculation around the corner of the waveform. This is why detailed analysis of complex waveforms can take a long time.

The ⟨final time value⟩ is the duration of the analysis. You must be sure that the time specified is long enough to ensure that all the data needed are calculated. All transient analyses begin at time equals zero and finish at the value specified in this part of the command.

The [⟨no step time⟩] may save some paper if all the data that are calculated at the beginning of the analysis are not important. The function of this command is to stop the printing of data to the output files for a specific amount of time. Printing of data to the output files starts at the time specified in this command.

The maximum step time limits the size of the maximum time step the program uses for calculation. The transient analysis will set a maximum step time of $1/50$ of the ⟨final time value⟩ if this parameter is not specified. For analyses that take a long time, this value may be too large to provide the necessary data for the corners of the waveform. If this is the case, you will usually see sharply angled portions of the waveforms generated by PROBE. When this occurs, you can include this part of the command.

There are two other command options listed, [/OP] and [UIC]. When a transient analysis is run, the dc bias point for the analysis is calculated before execution of the analysis. The bias point calculated for transient analysis may or may not be the same as the dc bias point for the circuit. The [/OP] command causes the data to be sent to the output file. These data are in a form similar to that of the regular bias point calculation. The [UIC] option allows initial conditions to be set for the circuit. If capacitors and inductors, or nodes in the circuit, have an initial condition specified, these conditions will be used in the bias point analysis. An example transient analysis line using all the parameters is

```
.TRAN/OP .1U 50U 10U 1U UIC
```

5.2. SIGNAL TYPES AVAILABLE IN PSPICE

Five signals are available in PSpice:

1. Sine wave
2. Pulse
3. Piecewise linear
4. Single-frequency FM
5. Exponential

These signals will only be active during the transient analysis portion of the program; that is, there may be the usual type of ac or dc analysis done using the .AC or .DC sweep. Then the transient analysis is done using the signal specified. To use any of the signals in an analysis we must discuss the parameters that define each type of signal. During the transient analysis these signals override any other input signals, just like the sweep voltage does during a sweep.

The method of writing the independent voltage or current source line is also important. It may contain all the types of source to be discussed, as well as the usual sources. An example of how to specify all the types of source on a single line is

```
V/I<name> <+ node> <- node> DC <value> AC <value> <magnitude>
+        <phase> <transient source type> (parameters)
```

Writing the source line this way is all right for analysis of simple circuits. Usually, the dc is the supply source and the ac and transient sources are at the input to the circuit and are on their own line or lines. Examples of the use of each of the signals will be given after all the signal types have been described.

5.3. THE SINE WAVE SIGNAL

We start with this type of signal source since it is the simplest of the types of signal that can be employed to analyze a circuit. It is the only pure signal in that it contains no harmonics; only the fundamental frequency is present. There are two basic ways in which this signal can be utilized. It can exist as a single signal with a single frequency at a constant amplitude or it can be exponentially damped.

There are six parameters that we can use to vary the sine wave in the manner we want:

Vo	Offset voltage
Va	Peak amplitude of the sine wave
freq	Frequency of the sine wave
td	Delay time before the sine wave starts
df	Exponential damping factor for the sine wave
phase	Phase of the sine wave, relative to zero phase

Four of the sine wave parameters have default values. This allows us to set the sine wave up the way that we want it to be. The syntax used to specify the sine wave is

```
V<name> <+ node> <- node> SIN( <vo> <va> <freq> <td> <df>
                        <phase> )
```

In all cases the minimum number of parameters that must be used to make a sine wave is three. The offset is a dc offset value and must always be specified, even if it is zero. The peak amplitude must be specified. Usually, the frequency of the sine wave will also be specified, but it also has an optional default. The default value is related to the final time set in the .TRAN command. The frequency will be set to 1/<final time value>. An example of a simple, constant-amplitude sine wave is

```
Vsin 1 0 SIN( 1V 10V 10KHz )
```

This sine wave has a 1 V dc offset, a peak amplitude of 10 V, and a frequency of 10kHz.

We should also talk about the three parameters that were allowed to default. The time delay, td, is a time that is allowed in the analysis before the sine wave starts. Time is always in seconds. The damping factor, df, is the exponent of the exponential e and has the form $\exp[-(\text{TIME} - \text{td})] \cdot df$, where df is the damping factor of the sine wave and is 1/s. The phase of the sine wave is set by using the specification "phase." You may set the phase to any value needed (between 0 and 360 degrees). The sine wave and its parameters are shown in Figure 5-1.

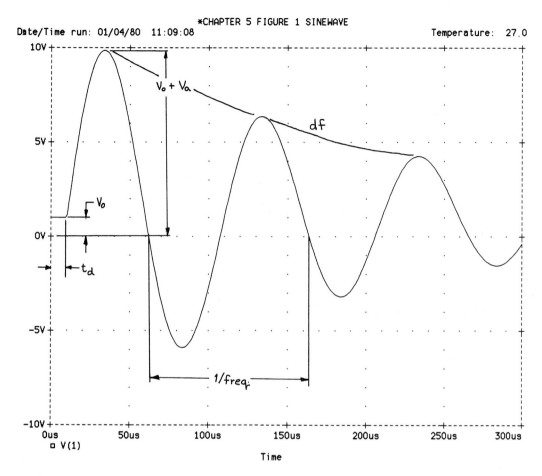

Figure 5-1

5.4. PULSE SIGNALS

The pulse signal is characterized by seven parameters that may be set by the designer:

v1	Starting value of voltage for the pulse
v2	Maximum voltage for the pulse
td	Delay time of the pulse
tr	Rise time of the pulse
tf	Fall time of the pulse
pw	Width of the pulse
per	Total period of the pulse

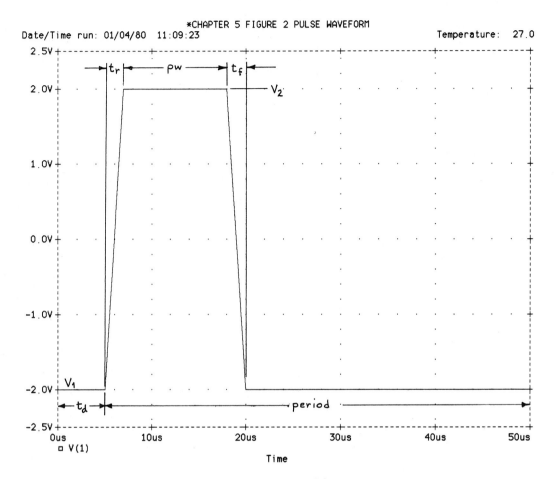

Figure 5-2

There are two parameters that do not have default values and must always be set: v1 and v2. All other values default either to zero or are time related to the .TRAN command. The rise and fall times will be equal to the maximum time step (1/50 of the final time), and the pulse width and period will be equal to the maximum (final) time. The syntax for specifying the pulse signal is

```
V<name> <node> <node> PULSE( <v1> <v2> <td> <tr> <tf> <pw>
                          <per> )
```

An example control line using a pulse source is

```
Vpulse 1 0 PULSE(-2 2 5U 2u 2u 11u 50u)
```

A pulse signal and its parameters are shown in Figure 5-2.

5.5. PIECEWISE LINEAR WAVEFORMS

This is a very useful function in that it allows us to simulate any type of waveform that can be made from straight-line segments (we could approximate a curve if the time and voltage steps are small enough). An example is the staircase waveform. There is one caveat to be stated here. The number of lines of code for a piecewise linear waveform can become many, very quickly, if the waveform is complex enough. This is not a problem, but it is time consuming to type in.

There are no default values for this type of waveform; all of the voltages and times must be specified. You may have as many as needed to complete the waveform. Each of the times specified is a corner or turning point of the waveform. The voltage levels are linearly interpolated between any two time points. The syntax for this type of waveform is

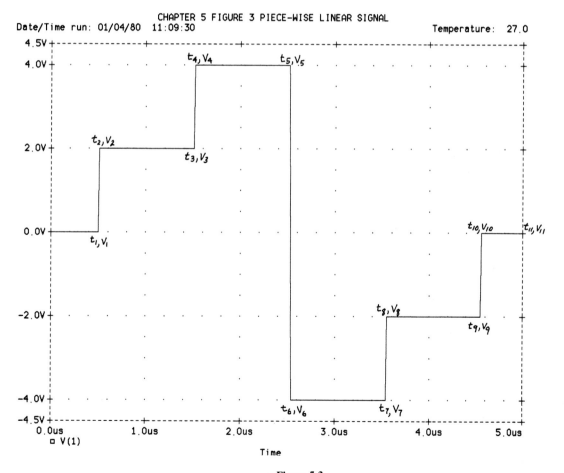

CHAPTER 5 FIGURE 3 PIECE-WISE LINEAR SIGNAL

Figure 5-3

```
V<name> <node> <node> PWL( <t1> <v1> <t2> <v2> · · · <tn>
                                        <vn> )
```

An example current source line for a piecewise linear waveform is

```
Vpwl 1 0 PWL(0 0V 0.01US 2V 1US 2V 1.01US 4V 2US 4V 2.02US
+        -4V 3US -4V 3.01US -2V 4US -2V 4.01US 0V)
```

Let us us examine the sample line to see the functioning of this type of signal. The signal starts at zero time and ends at 4.01 μs. Each of the time values in between represent a corner of the waveform. Thus the voltage rises from zero volts at zero time to 2 V in 0.01 μs. The voltage remains at 2 V until time reaches 1 μs. Between 1 and 1.01 μs, the voltage rises to 4 V, where it remains until 2 μs of time has elapsed. At this time the voltage falls to -4 V in 0.02 μs. The process continues until the waveform ends at 4.01 μs.

You can see that if a waveform is fairly complex, the netlist file can contain several lines of data for the signal, or perhaps more. An example piecewise linear waveform is shown in Figure 5-3.

5.6. THE SINGLE-FREQUENCY FM SIGNAL

This is a standard FM signal. Five parameters are used to set this signal up:

vo	Offset voltage of the carrier
va	Peak amplitude of the carrier voltage
fc	Carrier frequency
mdi	FM modulation index
fs	Modulation frequency

Of these parameters, the offset voltage and the peak amplitude of the carrier voltage must be specified. The other parameters will default to either zero or a value that is time related to the .TRAN command. The modulation index will be 0 when not specified and the modulation frequency and carrier frequency will be 1/max time value. The syntax for this signal is

```
V<name> <node> <node> SFFM( <vo> <va> <fc> <mdi> <fs> )
```

where SFFM represents single-frequency FM. A sample control line for this waveform is

```
Vsffm 1 0 SFFM(2 1 250KHz 5 25000Hz)
```

An example SFFM signal is shown in Figure 5-4.

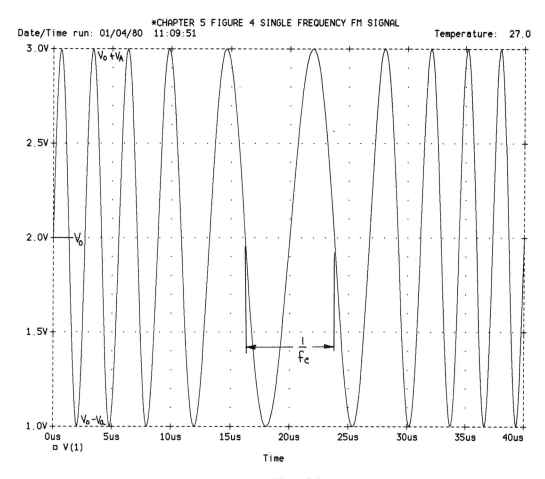

Figure 5-4

5.7. THE EXPONENTIAL SIGNAL

This is a source whose rise and fall times are exponential. There are six parameters to be assigned to this signal:

v1	Initial voltage of the exponential
v2	Peak voltage reached by the exponential
td1	Rise delay time
tc1	Rise time, time constant
td2	Fall delay time
tc2	Fall time, time constant

Of these constants, the initial voltage and the peak voltage must be specified. All of the other constants will default either to zero or to a value set by the .TRAN command. tc1 and tc2 will be equal to the time step set for the transient analysis. td2 will be equal to the time step and have any delay time specified added to it. The syntax for this source is

```
V<name> <node> <node> EXP( <v1> <v2> <td1> <tc1> <td2> <tc2> )
```

An example of specifying this source is

```
Vexp 1 0 EXP(0 10 .5msec 5msec 10msec 5msec)
```

An example of the exponential signal is shown in Figure 5-5.

The exponential signal bears a little more explanation here. When you use this signal, the fall delay time determines what voltage the exponential will reach. tc1 and tc2 are the time constants of the exponential rise and fall. If you apply 10 V, as in the sample line, and the fall delay time is equal to tc1, you will see the exponential rise to 0.632 of the applied voltage, 6.32 V. By the same token, the fall of the exponential will be to only 0.368 of the voltage of the rise, or, 2.35 V. If you want the exponential to rise beyond 6.32 V, the fall delay time needs to be long enough to assure that the signal reaches the desired value. If, for example, the fall delay time were twice tc1, the voltage would rise to 8.65 V.

The complete netlist for generating all of the signals is shown below. This is the simplest of netlists in that the signals are developed across only the signal source and a resistor. You can modify this netlist and the signal sources to see the effects of changing the parameters.

```
*CHAPTER 5 FIGURE 1 SINE WAVE
Vsin 1 0 SIN(1V 10V 10KHz 10U 5E3)
R1 1 0 10K
.TRAN 10U 300U
.PROBE
.END
*CHAPTER 5 FIGURE 2 PULSE WAVEFORM
Vpulse 1 0 PULSE (-2 2 5U 2u 2u 11u 50u)
R1 1 0 1E3
.TRAN 1U 50U
.PROBE
.END
*CHAPTER 5 FIGURE 3 PIECE-WISE LINEAR SIGNAL
Vpwl 1 0 PWL(0 0v 0.01u 2v 1u 2v 1.01u 4v 2u 4v 2.02u -4v
+     3u -4v 3.01u -2v 4u -2v 4.01u 0v)
R1 1 0 1E3
.TRAN .01U 5U
```

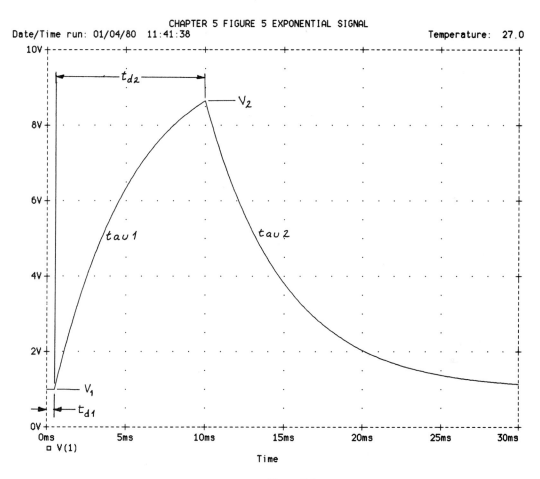

Figure 5-5

```
.PROBE
.END
*CHAPTER 5 FIGURE 4 SINGLE FREQUENCY FM SIGNAL
Vsffm 1 0 SFFM(2 1 250KHz 5 25000Hz)
R1 1 0 1E3
.TRAN .1U .04M
.PROBE
.END
*CHAPTER 5 FIGURE 5 EXPONENTIAL SIGNAL
Vexp 1 0 EXP(0 10 .5msec 5msec 10msec 5msec)
R1 1 0 1E3
.TRAN .1M 30M
.PROBE
.END
```

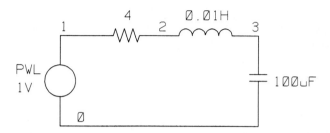

Figure 5-6

5.8. USING .TRAN FOR CIRCUIT ANALYSIS

To see how this type of analysis works, we will analyze a series LCR network. We will show the damped oscillation of this circuit when excited by a PWL pulse. The curve produced is for a circuit whose damping ratio is 0.2. The circuit for the filter is shown in Figure 5-6. The netlist for the network is

```
*DAMPING CHART FOR Z = 0.2
V1 1 0 PWL(0,0,1U,1,15.999M,1V,16M,0)
R1 1 2 4
L1 2 3 .01
C1 3 0 100E-6
.TRAN .005M 16M 0 0.1M
.PROBE
.END
```

Note from the netlist file that when the maximum step time is used, the no printing interval must be specified, even if it is zero. In this case the maximum step size is limited to 100 μs. This assures that good rounded corners will occur on the output waveform.

When this file is run, the output should be the same as shown in Figure 5-7. This output graph could also be normalized. It then can be used for any circuit whose damping ratio (ζ) is 0.2. The curves for any damping ratio can be graphed by calculating the resistance for the desired damping. The formula for the damping is

$$R = 2\zeta \sqrt{\frac{L}{C}}$$

where ζ is the damping ratio desired.

Another example of the use of transient analysis signals is one to produce an amplitude-modulated waveform. The device we will use to produce this waveform is a controlled source multiplier. In this case we will use a VCVS device. The output of the multiplier will contain the carrier and the sum and difference (sidebands) of the carrier and the two frequencies. When a 1-V peak, 50-kHz carrier, and a 0.9-V peak, 2-kHz modulating frequency are multiplied, an output voltage that varies between 1.9

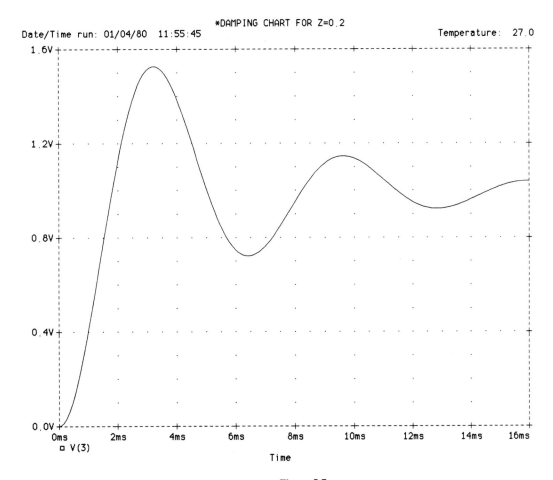

Figure 5-7

V at the peak and 0.1 V minimum is produced. This gives us a modulation factor, *m*, of 0.9. The netlist for this analysis is

```
*AMPLITUDE MODULATED WAVE - CHAPTER 5 FIGURE 8
V1 1 0 SIN(0 1 50K)
V2 2 0 SIN(1 .9 1K)
R1 1 0 1E9
R2 2 0 1E9
EMOD 3 0 POLY(2) (1,0) (2,0) 0 0 0 0 1
RE 3 0 1E6
.TRAN 10U 1.5M
.PROBE
.END
```

The output of the circuit is shown in Figure 5-8.

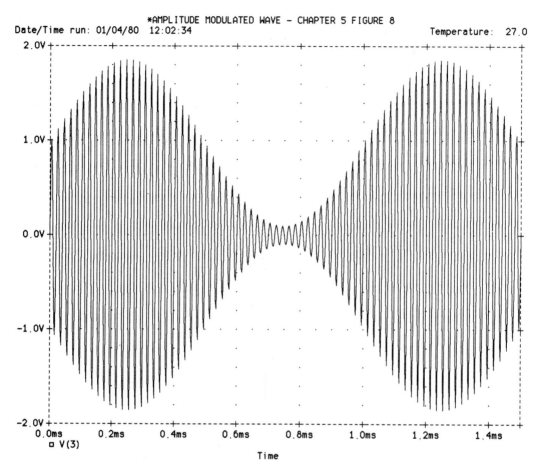

Figure 5-8

5.9. TRANSFORMERS IN PSPICE

The logical extension of the inductor is to add windings to the inductor, which couples them together, making a transformer. When a transformer is made, there are more parameters needed than for a simple inductor. Transformers in PSpice are handled using an inductor coupling coefficient command, K. K is the same as the coupling coefficient, k, used with any transformer. It is the percentage of the number of lines of flux generated in the primary of the transformer that link to the turns of wire in the secondary. If the windings are far apart, the value of k approaches 0; if they are very close together, k approaches 1.

The transformer may be tuned or untuned, and may have two or more windings. The general form for specifying coupling for a transformer is

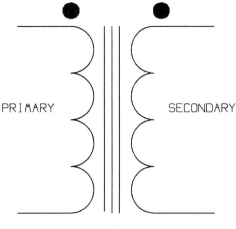

Figure 5-9

```
K<name> L<name> L<name>  ·  ·  · <coupling value>
K<name> L<name> L<name>  ·  ·  · <coupling value> <model name>
+ [size value]
```

K indicates to PSpice that two or more inductors will be coupled together. The inductors use the dot convention for polarity. This means that the polarity of the inductors is determined by the order of the nodes of the individual inductors, not the order in which they are listed. For this reason, the dot nodes must be carefully observed so that the polarity of the voltage induced is correct for each of the windings. The convention for the dotted nodes is shown in Figure 5-9. Connected as shown, the polarity of the current in the secondary inductor will be opposite that of the current in the primary.

The ⟨coupling value⟩ is the coefficient of mutual coupling for the transformer. This is a value between 0 and 1 and may not exceed 1 in any design. If more than one secondary winding is used, the coupling can be specified for all the windings using a single line. The line includes all the windings and the coupling value on that line is for all the windings.

[size value] is used when we are using a core model. It gives the number of laminations that you wish to have in a core. This value defaults to 1 if it is not specified. This parameter allows a single model to have as many laminations in a core as necessary. Nonlinear core models will be discussed shortly.

An example of the use of transformers is a center-tapped secondary transformer for a single full-wave rectifier. The input voltage to the primary is 170 V peak (120 V rms) at 60 Hz. The output voltage is 17 V peak (12 V rms) per secondary half. The filter capacitor has been left off to show the rectified output. The resistances of the windings have also been left out. Only a 1-Ω resistor has been added in series with LPRI. This prevents a voltage loop that will develop if LPRI is connected directly to the ac source.

Before proceeding with the netlist for the circuit, we need to mention some of the factors that will affect the transformer. Since this is a full-wave center-tapped transformer, we need to determine inductances for the windings to have the proper output voltage from the rectifier. For this transformer, we consider the turns ratio to be that of one-half of the total secondary. The inductance for each secondary is the primary inductance divided by the square of the turns ratio. Conversely, if we know the secondary inductance, we can find the primary inductance by multiplying the secondary inductance by the square of the turns ratio. In this case the turns ratio is 10:1.

We can assume some secondary inductance, or we can calculate it if we know the power factor of the transformer. In this case we will assume a secondary inductance of 0.015 H. This gives a primary inductance of 1.5 H. We will need two secondary inductances to make up the total secondary. Simple calculation for the output of the rectifier gives the output peak voltage to be 17 V across the load resistor. The diodes

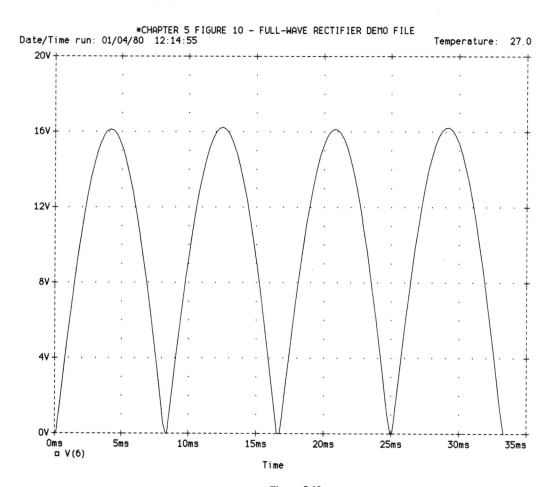

Figure 5-10

will probably drop more than 0.7 V, so the transformer primary or secondary inductance can be adjusted to compensate for this. The diodes chosen for this analysis are the 1N4001 power diodes in the library. The netlist for this analysis is

```
*FULL-WAVE RECTIFIER DEMO FILE
VAC 1 0 SIN(0 170 60)
RAC 1 2 1
LPRI 2 0 1.5                       ;transformer primary
LSEC1 3 4 .015                     ;secondary 1
LSEC2 4 5 0.15                     ;secondary 2
KTRAN LPRI LSEC1 LSEC2 0.999       ;coupling for transformer.
R2 4 0 1E9
D1 3 6 D1N4001
D2 5 6 D1N4001
R3 4 6 1K
.LIB
.TRAN .1M 33.3333M 0 .1M
.PROBE
.END
```

The output of this file is shown in Figure 5-10. This file will also be analyzed when we do the Fourier analysis part of the transient command and the average dc output voltage will be found to be close to the value of $0.618V_{pk}$ predicted by calculation.

5.9.1. Nonlinear Magnetic Core Models

When we design inductors and transformers that are wound on a core, it is important that the core have the proper characteristics. The characteristics are initial permeability, saturation, hysteresis, and core losses. PSpice magnetic core models include all of these effects. The model used by PSpice is the Jiles–Atherton model (Jiles and Atherton, 1986). This model produces the $B–H$ curve of the core being investigated. Unfortunately, this type of modeling also requires a large amount of PFM (pure flaming magic) to determine a proper $B–H$ curve for the core being used.

When the model name is included in the transformer control line, there are four parameters that change in the transformer:

1. The mutual coupling inductor is now a nonlinear magnetic core device.
2. The individual inductors are now transformer windings rather than inductors, and the number that formerly was the inductance is now the number of turns of the winding.
3. The former list of coupled inductors may be just one inductor.
4. You must have a .MODEL statement to model the core device.

There are nine parameters of a core model that may be changed (Table 5-1).

TABLE 5-1

Name	Parameter	Default	Unit
AREA	Core magnetic cross section	0.1	cm^2
PATH	Core magnetic path length	1.0	cm
GAP	Effective air-gap width	0	cm
PACK	Number of laminations	1.0	
MS	Saturation magnetization	1E+6	ampere/meter
ALPHA	Field parameter	1E−3	
A	Shape parameter (squareness)	1E+3	ampere/meter
C	Domain wall flexing constant	0.2	
K	Domain wall pinning constant	500	

The first four parameters are geometric constants of the core being used and the number of cores that are stacked together. The geometric constants are in MKS units. The first three are obtained from the manufacturer's data sheet for the core. The fourth you must know. The last five parameters are constants that determine the magnetic operation of the core.

A discussion of the last six parameters of the core follows.

PACK When the core is defined, it is a single lamination. The core may contain more than one lamination. The number of laminations may be specified using this value, and a single model written for the transformer. The default for this parameter is 1.

MS Magnetic saturation current for the core of the transformer. A first approximation for this value can be obtained by using the formula

$$MS = \frac{B}{0.01257}$$

where B is the maximum flux density in gauss. The default value for this value is 1E + 6 Amperes per meter.

ALPHA Mean field parameter. Determined experimentally, the default for this parameter is 0.001. Start with a value of zero for this value and increase it gradually. It can have a large effect on the squareness of the curve.

A Shape parameter for the curve. Decreasing its value raises the knee of the hysteresis curve, affecting the squareness of the curve also. Increasing this value causes the knee to curve more gently, but also causes B to decrease.

C Magnetic domain wall flexing constant. It affects how broad the separation of the hysteresis curve is.

K Domain wall pinning constant. It affects how wide the curve is. Use this to get the $B = 0$ crossing points of the curve just right.

To see how the various parameters affect the core, let us do a core analysis. We will use a current sine wave input to a coil on the core. Then we will display the hysteresis curve of the core being analyzed. Much of the core model work will be experimental. We will have to play around with some of the values to get the curve "just right." For this demonstration we will trace the hysteresis curve for a core that is used for power converters and similar low-frequency applications.

An example of the hysteresis curve of the core we will try to duplicate is shown in Figure 5-11. The geometric parameters of the core are

$$\text{AREA} = 0.403 \text{ cm}^2$$
$$\text{PATH} = 6.27 \text{ cm}$$
$$\text{VOLUME} = 2.52 \text{ cm}^3$$

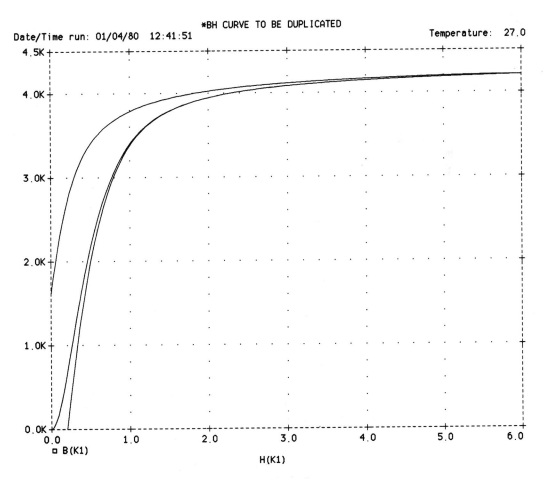

Figure 5-11

The specifications for this core are

$$B_{max} = 4200 \text{ G}$$
$$H_{max} = 6 \text{ Oe}$$

We can see from the diagram supplied that the hysteresis curve crosses the axis of the B–H curve at approximately ± 0.2 Oe. This is important to know so that the values of C and K can be made to be correct.

The first parameter to determine is the saturation current of the core. Then we will run a preliminary curve for the core. For now A, C, and K can be allowed to default, and Alpha is set to zero. The calculated value of MS is

$$MS = \frac{4200}{0.01257} = 3.3413 \times 10^5 \text{ A}$$

In PSpice core models have the designation K. K can be followed by a name if desired. The core we will use is one from the Fair-Rite Corporation catalog. Its part number is 59-77-001401. The number 77 is the mix material of the core. This core mix is recommended for use in power conversion transformers, power-line filters, and ignition coils. We will give the core model the name K77CORE.

We will try to get our core model to have the same shape as the BH curve in Figure 5-11. The preliminary netlist for the core is

```
*BH DEMO FILE
I1  0  1  SIN(0 1 1)
R1  1  0  1
L1  1  0  30
K1  L1  0.9999  K77CORE     ;core model K77CORE
.MODEL K77CORE CORE(MS=3.3413E5 AREA=0.403 PATH=6.27
+       ALPHA=0)
.OPTIONS ITL5 = 0
.TRAN 0.01 1.25
.PROBE
.END
```

The output curve of this file is shown in Figure 5-12. Not what we wanted but we can get information from this curve that will allow us to get the model much closer on the next try. If we analyze this curve we can set A and K to closer values. As stated before, A changes the shape of the knee of the curve and also lowers the flux density of the coil. From these two values we find that A is too large. The maximum flux density is only around 300 G. As a first guess, we must decrease A by more than 20 times to get the flux density to near 4200 G. K causes the curve to widen. The curve now crosses the $B = 0$ axis at almost 3 Oe, so the value of K is also too large. We need to lower this value by at least 10 times. The default for A is 1E + 3. The default

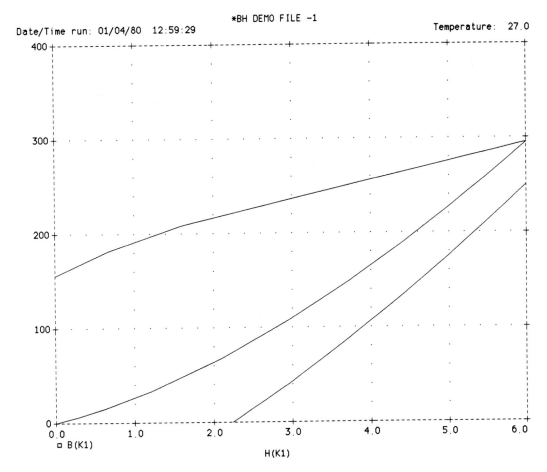

Figure 5-12

for K is 500. Lowering each of these values to 50 should give us a much better curve. The core model line should now be changed to

```
.MODEL K77CORE CORE(MS=3.3413E5 AREA=0.403 PATH=6.27
+       ALPHA=0 A=50 K=50)
```

Let us rerun the curve using the new values and see how the curve changes. The curve is shown in Figure 5-13. Certainly, this is better than the curve that we first had. The values of A and K are still not correct but we can lower them until we get the right curve crossings and shape of the knee. The curve now crosses the $B = 0$ axis at about 0.5, so we could cut the value of K in half to see the effect. The maximum flux density

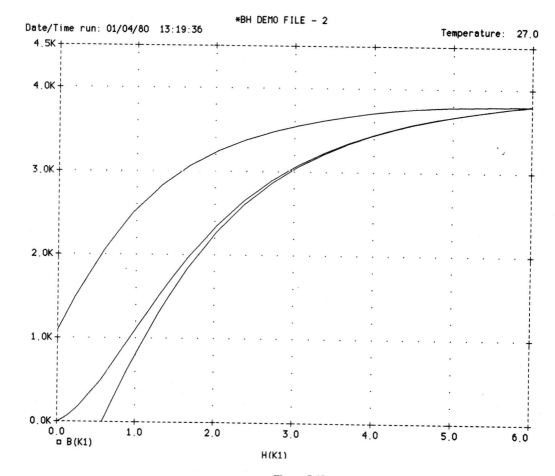

Figure 5-13

is still too low also, so we can cut the value of A in half also. The curve produced by doing this is shown in Figure 5-14. Although this curve is much closer to the desired curve, it is still not right. Some final trimming is needed to make it the required shape. We can now start setting the values of Alpha and C. Alpha raises the knee of the curve and must be done carefully. It has a profound effect on the shape of the curve. If we were to let Alpha default, the curve would be very wide and would assume an almost square shape. For this reason the value of Alpha must be much smaller than the default value. If C is allowed to default, it will narrow the curve at the zero crossing points slightly. We can then use C to trim the curve crossing to the exact crossing points required.

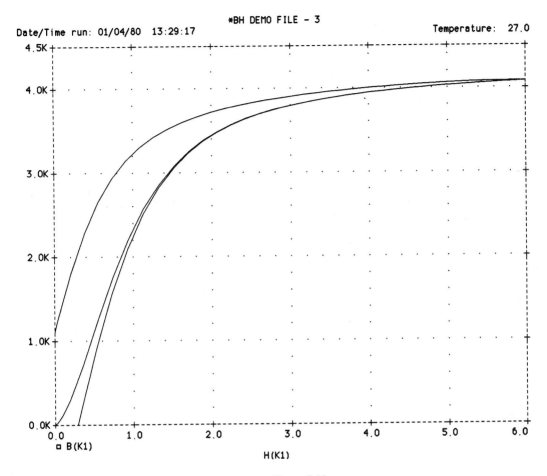

Figure 5-14

This part of the process will require that you analyze the curve several times until you get the curve correct. For the final curve, the value of Alpha was found to be $1E - 6$ and the value of C was found to be 0.01. The correct model command line then is

```
.MODEL K77CORE CORE(MS=3.3413E5 AREA=0.403 PATH=6.27
+        ALPHA=1E-6 C=0.1 A=12 K=18)
```

The curve then has the shape shown in Figure 5-15, and this is close enough to the published curve for any use we may have.

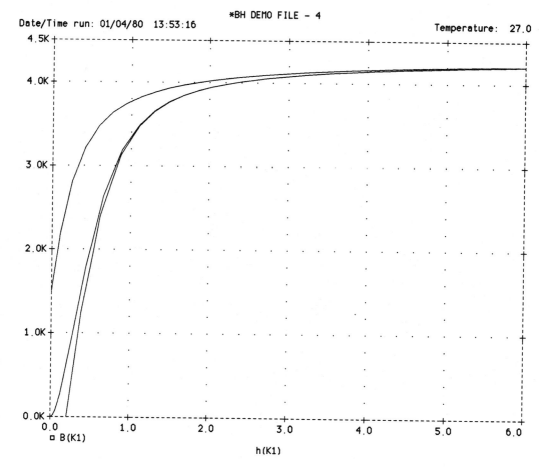

Date/Time run: 01/04/80 13:53:16

Temperature: 27.0

Figure 5-15

5.10. .FOUR FOR HARMONIC COMPOSITION

It is well known that any waveform other than the simple sine wave is composed of several components. These components are an average (or dc) value, a sinusoidal wave at the fundamental frequency, and an infinite number of harmonic sinusoids that are time related to the fundamental. This group of sinusoids is called the spectrum of the waveform. One of the uses of the spectrum of a wave is to determine the bandwidth required to transmit a complex waveform.

The type of analysis used to analyze the complex waveform is the analysis first done by J. B. J. Fourier in 1812. This analysis disassembles the complex wave into its fundamental and harmonics. This analysis is included in PSpice and in SPICE. It is limited to the first nine harmonics, including the fundamental frequency. The type of analysis used is the discrete Fourier transform (DFT). The general syntax for the Fourier analysis in PSpice is

```
.FOUR <frequency> <output> · · ·
```

where ⟨frequency⟩ is the fundamental frequency, 1/period of the wave being analyzed. ⟨output⟩ is the node or device for which the measurement is made.

Fourier analysis is done only during a transient analysis. It therefore must be in a netlist with the .TRAN command. There can only be one .FOUR command in any netlist, but you may specify as many output nodes or devices as necessary. Also, this type of analysis is automatically printed to the output file so there is no need to use the .PRINT, .PLOT, or .PROBE command to save the values. The analysis is done at the end of the transient analysis backwards one period (final time value − period). You have the option to run the analysis for one period or for several. This depends on whether or not the system has a transient residue at the start of the analysis. In any case, it is a good idea to run the analysis for several cycles.

Let's do a netlist to show the use of the .FOUR command. We will then look at the output file. The netlist we shall use is the full-wave rectifier that we developed earlier. The netlist for this analysis is repeated here for clarity.

```
*FULL-WAVE RECTIFIER DEMO FILE
VAC 1 0 SIN(0 170 60)
RAC 1 2 1
LPRI 2 0 1
LSEC1 3 4 .015
LSEC2 4 5 .015
KTRAN LPRI LSEC1 LSEC2 0.999
R2 4 0 1E9
D1 3 6 D1N4001
D2 5 6 D1N4001
R3 4 6 1K
.LIB
.TRAN .1M 33.3333M 0 .1M
.FOUR 60 V(6)
.END
```

The voltage output of this netlist is the same as the original netlist. The Fourier analysis portion of the output file is shown in Analysis 5-1.

When we examine the Fourier components chart we find that the dc component of the output is 10.04 V. This is about 0.617 V_p and is in close agreement with the ideal of 0.637 V_p. The chart gives four values for the fundamental and harmonics: first, the Fourier component in volts; second, the Fourier component voltage normalized to the fundamental component voltage; third, the phase of the component in degrees; and fourth, the phase component normalized to zero degrees.

The fundamental frequency used in this analysis is 60 Hz. This is the frequency

```
******* 10/13/90 ******* Evaluation PSpice (January 1989) ******* 10:04:21 *******

*CHAPTER 5 FIGURE 10 - FULL-WAVE RECTIFIER DEMO FILE ANALYSIS - 5-1

****      CIRCUIT DESCRIPTION


***********************************************************************
VAC 1 0 SIN(0 170 60)
RAC 1 2 1
LPRI 2 0 1.36
LSEC1 3 4 .015
LSEC2 4 5 .015
KTRAN LPRI LSEC1 LSEC2 0.999
R2 4 0 1E9
D1 3 6 D1N4001
D2 5 6 D1N4001
R3 4 6 1K
.LIB
.FOUR 60 V(6)
.TRAN .1M 33.3333M 0 .1M
.END

****      Diode MODEL PARAMETERS


          D1N4001
   IS     1.923000E-15
   RS     .08997
   CJO    1.000000E-12


****      INITIAL TRANSIENT SOLUTION        TEMPERATURE =    27.000 DEG C


NODE   VOLTAGE      NODE   VOLTAGE      NODE   VOLTAGE      NODE   VOLTAGE

(   1)    0.0000  (    2)    0.0000  (    3)    0.0000  (    4)    0.0000

(   5)    0.0000  (    6)-2.053E-21


    VOLTAGE SOURCE CURRENTS
    NAME         CURRENT

    VAC          0.000E+00

    TOTAL POWER DISSIPATION   0.00E+00  WATTS
```

Analysis 5-1

analyzed at the output also. The first line shows that the 60-Hz component is very small in comparison to 120 Hz and the other seven components. This is as it should be. The next seven lines are the other harmonics of the fundamental. The total harmonic distortion is calculated by summing the squares of harmonics 2 through 9, then taking the square root of this number and dividing it by the fundamental voltage and multiplying

```
****      FOURIER ANALYSIS                 TEMPERATURE =   27.000 DEG C
```

FOURIER COMPONENTS OF TRANSIENT RESPONSE V(6)

DC COMPONENT =   1.060968E+01

HARMONIC NO	FREQUENCY (HZ)	FOURIER COMPONENT	NORMALIZED COMPONENT	PHASE (DEG)	NORMALIZED PHASE (DEG)
1	6.000E+01	4.267E-02	1.000E+00	-1.795E+02	0.000E+00
2	1.200E+02	7.522E+00	1.763E+02	-8.978E+01	8.970E+01
3	1.800E+02	1.448E-02	3.394E-01	-1.781E+02	1.406E+00
4	2.400E+02	1.481E+00	3.471E+01	-8.958E+01	8.991E+01
5	3.000E+02	8.384E-03	1.965E-01	-1.786E+02	9.006E-01
6	3.600E+02	6.210E-01	1.455E+01	-8.937E+01	9.011E+01
7	4.200E+02	5.894E-03	1.381E-01	-1.757E+02	3.819E+00
8	4.800E+02	3.355E-01	7.863E+00	-8.921E+01	9.027E+01
9	5.400E+02	4.389E-03	1.029E-01	-1.766E+02	2.844E+00

TOTAL HARMONIC DISTORTION =   1.804112E+04 PERCENT

JOB CONCLUDED

TOTAL JOB TIME   44.16

Analysis 5-1 (cont'd)

by 100. This gives a very large distortion, greater than 10,000%. This is because the input frequency is 60 Hz and the output frequency is 120 Hz. If we were to have analyzed the output at 120 Hz for the fundamental, the distortion would be about 22%.

Fourier analysis using PROBE   PROBE has a Fourier transform function also. We shall use the same full-wave rectifier as before to examine the spectrum generated by this analysis using PROBE. The netlist with PROBE is

```
*FULL-WAVE RECTIFIER DEMO FILE
VAC 1 0 SIN(0 170 60)
RAC 1 2 1
LPRI 2 0 1
LSEC1 3 4 .015
LSEC2 4 5 .015
KTRAN LPRI LSEC1 LSEC2 0.999
R2 4 0 1E9
D1 3 6 D1N4001
D2 5 6 D1N4001
R3 4 6 1K
.LIB
.TRAN .1M 33.3333M 0 .1M
.PROBE
.END
```

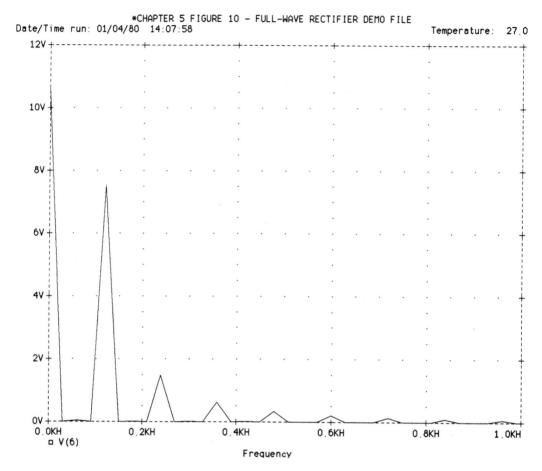

Figure 5-16

First, get the full-wave rectified output on the PROBE screen. We can then view the Fourier waveform by using the *x*-axis menu (type 3, then 6 for the student version; type x for *x* axis, then type F for Fourier, for the advanced version). The analysis of the output is shown in Figure 5-16. You will also need to expand the *x* axis to see the spikes and their widths clearly.

The type of transform that is used is also the DFT. This is made to be the fast Fourier transform (FFT) by making the highest point calculated a power of 2, then dividing this number by the number of calculation points to produce evenly spaced points. The points are then calculated and the values scaled to bring them in line with the values calculated for the output table.

The resolution of the waveform is also important in that the spikes formed in the output graph can be quite wide. This is a function of the Fourier transform itself. The resolution is the reciprocal of the period of the original waveform. For our analysis the resolution is then 1/0.01667, or 60 Hz. To make the output spikes narrower, we must make the analysis run longer so that more points are generated. You can try making the analysis run for 5 cycles of the input, or 0.08333 s. This should narrow the spikes at the harmonic frequencies.

The range of frequency covered by the x axis in a Fourier analysis is related to the number of calculated points in the analysis. It is also related to the period of the frequency being analyzed. If our analysis gave us 64 output points, the frequency range would be 64 × 60 or 3840 Hz. This is the frequency range on the graph for this analysis. Understand that if the input frequency were higher but analysis time were not changed, the number of calculated points would be larger. The spikes would then be narrower and the frequency range would be wider.

5.11. .IC, UIC, AND .NODESET FOR INITIAL CONDITIONS

Up to now we have not had a need to preset node conditions for an analysis to be done. In certain analyses, some of the nodes must have conditions specified before circuit analysis is started. An example of this might be to preset a digital circuit to a desired state. To do this, PSpice has three commands that can be used: they are IC, .IC and .NODESET. .IC is active only during transient analysis, and then only if the UIC command is present. .NODESET is active for both ac and dc sweep analysis as well as transient analysis. If both a .NODESET and a .IC command are contained in a transient analysis, .IC overrides the .NODESET command. IC for devices, capacitors, and inductors is related to these commands but is contained in the device line itself. This command is also active only during a transient analysis. If both are present in a single netlist, it is possible that we could try to force two different voltages across the same component. This will cause a problem in the analysis. Care must be taken if both .IC and IC are to be used together. Do not name the nodes of a device with an IC specification in a .IC line.

When the .IC command is used, power supplies with an internal resistance of 2 mΩ (to prevent voltage loops) are placed at the nodes specified. This forces the nodes to be at the voltages specified by .IC during the bias point calculation. During the bias point calculation the nodes remain at the voltages set by .IC. Once transient analysis starts, the power supplies are released and the nodes are free to change as the analysis continues. This command affects only the transient bias point calculation, not the dc bias point calculation. Any dc bias points or sweeps will be conducted in the usual manner.

.NODESET does not work the same as the .IC command. .NODESET sets the nodes to the voltage specified for the first part of the bias point calculation. This is done by placing power supplies with 2 mΩ resistance in series with the nodes. The nodes are then released from the initial bias point and the final bias point is found. After the final bias point calculation, the nodes may not be at the voltage specified by .NODESET. You can see that this type of node setting provides only a preliminary guess for the voltages at the nodes. This command is also effective during the regular bias point calculation, not just a transient calculation. If an .IC command is also present for the same nodes as .NODESET, it will override the .NODESET command for the transient analysis.

.NODESET also has an effect on dc sweeps. The first step of the sweep uses the value set by .NODESET. All other steps of the sweep ignore this value. If the sweep is a nested sweep, only the first step of the inner sweep is affected by .NODESET.

SUMMARY

PSpice can do analysis over a specific time period, as well as frequency response analysis. To do an analysis based on time, the command .TRAN must be used.

When a time analysis is done, the user must supply an input signal that is one of five types available. The signal types are the sine wave, pulse, exponential, piecewise linear, and single-frequency FM.

Four of the signal types have several parameters that can be specified, and some of these can be allowed to default. The piecewise linear signal has no default values.

When a transient analysis is done, the circuit devices are no longer treated as linear devices capable of any output voltage or current.

When a transformer is needed, the coupling coefficient command, K, is used to couple the windings for the transformer together. K can never be equal to or greater than 1.

Nonlinear magnetic core models can be made for PSpice, by changing the parameters for the magnetic core model.

A Fourier decomposition of a signal can be done to show the relative amount of harmonic distortion in the signal.

SELF-EVALUATION

1. Using a piecewise linear waveform, make a LCR circuit with a damping of 0.15 and plot the output waveform using PROBE. The circuit constants are $L = 10 \ \mu H$ and $C = 100$ pF. You are to choose R using the formula given for ζ.

2. Using the exponential waveform, make a waveform that rises to 86.5% of 30 V in 10 ms and then falls to zero in 5 ms. Plot the output using PROBE.

3. Using two sine waves of 1 V amplitude, 0 offset voltage, 0 damping, 1 kHz frequency, 0 delay, and 90° phase difference between the two, make a circuit that sums the difference of the two sine waves as its output. Plot the output using PROBE.

4. Using the full-wave rectifier netlist from this chapter, add a smoothing capacitor of 2200 μF, and change the load resistor to 25 Ω. Determine the ripple voltage for this load. Using the plot control function of PROBE, use separate graphs to show the peak currents in the load resistor, diodes, and smoothing capacitor.

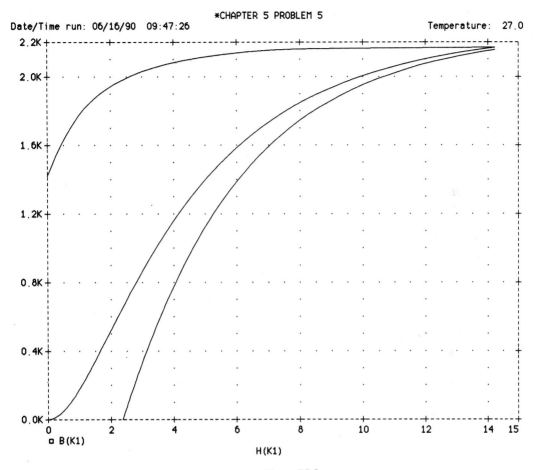

Figure P5-5

5. A sample hysteresis curve of a magnetic core is shown in Figure P5-5. The core material is number 65. This material is recommended for broadband and IF transformer cores for CATV and linear amplifiers. You are to make a model for this core material. The geometric core parameters are

$$\text{AREA} = 0.403 \text{ cm}^2$$
$$\text{PATH} = 6.27 \text{ cm}$$
$$\text{VOLUME} = 2.52 \text{ cm}^3$$

(You may care to store this model in your magnetic device library when you are done.)

6. Do a Fourier analysis of the amplitude modulator made earlier in the netlist. Check to see that the sidebands reflect the voltage that should occur with 90% modulation.

6

Integrated Circuit Models

OBJECTIVES

1. To learn methods of specifying operational amplifiers in PSpice.
2. To learn how to analyze operational amplifier circuits of several different types.
3. To learn to use the ideal operational amplifiers with the active operational amplifiers.

INTRODUCTION

The operational amplifier integrated circuit is probably the most versatile of the devices that we have today. The range of circuit types that can be implemented using the op amp is too large to be iterated here. PSpice has op amps and comparators in its libraries. If we need to use one that is not in the library, we should have an idea of how an op amp is modeled. We have already modeled ideal op amps as subcircuits, but these models will not perform the same as an actual model.

There are other types of integrated circuits that are not op amps that are used to perform different functions. We shall examine a few of these and make subcircuit netlists for some of them. We can store the netlists in a library we create.

6.1. OPERATIONAL AMPLIFIER MODELS

In PSpice, there are no op amp models as such. Rather, op amp models are made using subcircuits. This type of device is often called a *macromodel*. The subcircuits are called using the X⟨name⟩ command. The same rules apply for using the op amp as apply for using the subcircuits that we studied in Chapter 4. The op amps are contained in the libraries in PSpice. Since there are five connections to most op amps, there are five connections from the outside world as well. The op amp connections that we must specify are:

1. Noninverting input
2. Inverting input
3. Positive voltage supply
4. Negative voltage supply
5. Output

The order shown is also the order in which the connections for the op amp must be specified in the subcircuit call.

The model used for op amps is one that was devised in 1974 by G. R. Boyle and co-workers (Boyle et al., 1974). This model is accurate and the design equations for developing the op amp model are given in the article. The model for an op amp subcircuit is shown in Figure 6-1. Notice that the op amp subcircuit uses only two transistors and four diodes. The two transistors form the differential input amplifier of the op

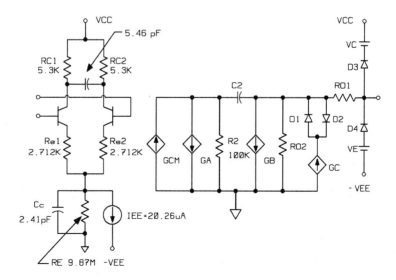

Figure 6-1

amp. The four diodes are used to model the effects of saturation. The rest of the devices are either dependent or independent sources. Four of the devices are voltage-controlled current sources, G devices. The emitter current source is an independent source. Making the model is not difficult once we have the necessary parameters. An example using the LM149 op amp model is

```
CONNECTIONS      NONINVERTING INPUT
*                 ¦  INVERTING INPUT
*                 ¦  ¦  POSITIVE SUPPLY
*                 ¦  ¦  ¦  NEGATIVE SUPPLY
*                 ¦  ¦  ¦  ¦  OUTPUT
*                 ¦  ¦  ¦  ¦
.SUBCKT LM149 40 30 10 20 120
R1 10 50 5.3K
R2 70 90 2.712K
R3 10 60 5.3K
R4 80 90 2.712K
R5 90 0 9.87MEG
C1 50 60 5.46P
C2 90 0 2.41P
Q1 60 30 80 QMOD2
Q2 50 40 70 QMOD1
IBIAS 90 20 10U
GCM 0 100 90 0 5.9N
GGAIN 100 0 60 50 150U
RGAIN 100 0 100K
GAMP 105 0 100 0 247.5M
RAMP 105 0 42.87K
CC 105 100 3E - 12
D1 105 115 DMOD
D2 115 105 DMOD
GOUT 0 115 120 0 46.96
ROUT 115 0 21.3M
RX 105 120 32
D3 120 125 DMOD
D4 130 120 DMOD
VC 10 125 1.803V
VE 130 20 2.803V
.MODEL QMOD1 NPN(IS=8E-16 BF=144)
.MODEL QMOD2 NPN(IS=8E-16 BF=112)
.MODEL DMOD D
.ENDS
```

Note also that the feedback capacitor C2 may be left out. The connections to this capacitor may be brought out to terminals on the op amp for compensation. This would give two additional terminals to the description of the op amp. To gain experience using the op amp models, lets analyze several circuits that are in common use.

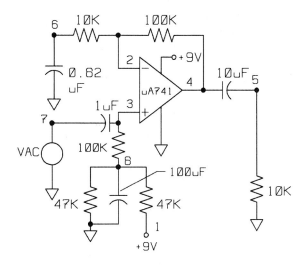

Figure 6-2

6.1.1. Audio Amplifier

The audio amplifier will have a low-frequency cutin at 20 Hz. We will run the circuit from a single 9-V supply. This will point out some of the differences between this and the ideal op amp made previously. The circuit for the amplifier is shown in Figure 6-2. The netlist for this circuit is

```
*LM741 OPAMP AUDIO AMPLIFIER
VCC 1 0 9VOLTS
VAC 7 0 AC 0.1VOLT
R1 1 8 47K
R2 8 0 47K
R3 2 6 10K
R4 2 4 100K
R5 5 0 10K
R6 3 8 100K
C1 6 0 .82U
C2 7 3 1U
C3 4 5 10U
C4 8 0 100U
XAMP 3 2 1 0 4 UA741
.LIB
.AC DEC 100 1 1E6
.PROBE
.END
```

It is also instructive to look at the dc output file of this netlist. The dc output file is shown in Analysis 6-1. Note that the output voltage at node 5 is approximately half of

```
******* 01/04/80 ******* Evaluation PSpice (January 1990) ******* 11:07:13 *******

*LM741 OPAMP AUDIO AMPLIFIER ANALYSIS 6-1

****        CIRCUIT DESCRIPTION

******************************************************************************
VCC 1 0 9VOLTS
VAC 7 0 AC 0.1VOLT
R1 1 8 47K
R2 8 0 47K
R3 2 6 10K
R4 2 4 100K
R5 5 0 10K
R6 3 8 100K
*RG 7 8 1K
C1 6 0 .82U
C2 7 3 1U
C3 4 5 10U
XAMP 3 2 1 0 4 UA741
.LIB
.AC DEC 25 1 1E6
.PROBE
.END

******* 01/04/80 ******* Evaluation PSpice (January 1990) ******* 11:07:13 *******

*LM741 OPAMP AUDIO AMPLIFIER

****        Diode MODEL PARAMETERS

******************************************************************************

            XAMP.dx
    IS    800.000000E-18
    RS       1

******* 01/04/80 ******* Evaluation PSpice (January 1990) ******* 11:07:13 *******

*LM741 OPAMP AUDIO AMPLIFIER

****        BJT MODEL PARAMETERS

******************************************************************************

            XAMP.qx
            NPN
    IS    800.000000E-18
    BF     93.75
    NF       1
    BR       1
    NR       1
```

Analysis 6-1

the supply voltage. This is expected for this type of circuit. The input resistance of this circuit is controlled by the parallel resistance at the input. Since the input resistance seen at the noninverting input is quite high, the input resistance is about equal to the resistor at the input.

```
******* 01/04/80 ******* Evaluation PSpice (January 1990) ******* 11:07:13 *******
*LM741 OPAMP AUDIO AMPLIFIER
****     SMALL SIGNAL BIAS SOLUTION        TEMPERATURE =   27.000 DEG C
**********************************************************************************

NODE    VOLTAGE       NODE    VOLTAGE     NODE    VOLTAGE     NODE    VOLTAGE

(    1)    9.0000   (    2)    4.4902  (     3)    4.4902  (    4)    4.4982

(    5)    0.0000   (    6)    4.4902  (     7)    0.0000  (    8)    4.4981

(XAMP.6) 173.3E-09 (XAMP.7)     4.4982 (XAMP.8)     4.4982 (XAMP.9)    0.0000

(XAMP.10)    3.8825              (XAMP.11)     8.9603

(XAMP.12)    8.9603              (XAMP.13)     3.8964

(XAMP.14)    3.8964              (XAMP.53)     8.0000

(XAMP.54)    1.0000              (XAMP.90) 78.21E-06

(XAMP.91)   40.0000              (XAMP.92)   -40.0000

(XAMP.99)    4.5000
```

```
        VOLTAGE SOURCE CURRENTS
        NAME            CURRENT

        VCC            -6.063E-04
        VAC             0.000E+00
        XAMP.vb         1.733E-12
        XAMP.vc         3.503E-12
        XAMP.ve         3.499E-12
        XAMP.vlim       7.821E-08
        XAMP.vlp       -4.000E-11
        XAMP.vln       -4.000E-11

        TOTAL POWER DISSIPATION   5.46E-03  WATTS

            JOB CONCLUDED

        TOTAL JOB TIME            34.33
```

Analysis 6-1 (cont'd)

6.1.2. Instrumentation Amplifier

Another widely used circuit is the instrumentation amplifier. This amplifier can be tested using two ideal op amps and a single active op amp. It is necessary to do this if you are using the student or evaluation version of PSpice. The active op amp shows saturation when the input range is exceeded. Ideal op amps do not saturate, so a transfer curve with limiting for the amplifying system cannot be obtained. Ideal op amps also provide a saving in the number of nodes required for the netlist. This amplifier has a gain of 1000 and is used to measure temperature with a RTD. The RTD is in a bridge configuration and the amplifier provides a linear output from $-40°C$ to $+40°C$. The

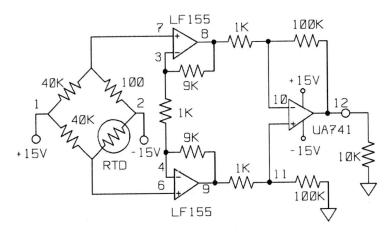

Figure 6-3

resistance of the RTD is 100 Ω at 0°C. The American standard, $\alpha = 0.00392$ Ω/Ω/°C is used to calculate the value of the RTD at any temperature. The value of the RTD resistance is adjusted for the nominal analysis temperature of 27°C used in PSpice. This value is 110.584 Ω. The RTD is a linear device, compared to thermocouples and thermistors. It does, however, suffer from self-heating; for this reason, the lower the current in the bridge, the better. When this circuit is run we will use a temperature sweep from −50 to +50°C. The resistance range of the RTD will then be from 78 to 121 Ω. The circuit of the instrumentation amplifier is shown in Figure 6-3.

The netlist for the amplifier is

```
*INSTRUMENTATION AMP DEMO
VCC 1 0 15
VEE 2 0 -15
RTD 6 2 RRTD 111.584
R1 1 7 40K
R2 7 2 100
R3 1 6 40K
R4 3 4 1K
R5 3 8 9K
R6 4 9 9K
R7 8 10 1K
R8 9 11 1K
R9 11 0 100
R10 10 12 100K
R11 12 0 10K
XPOS 7 3 8 LF155
XNEG 6 4 9 LF155
XOUT 11 10 1 2 12 UA741
.MODEL RRTD RES (R = 1 TC1 = .00392)
.LIB
```

```
.LIB SUB.LIB
.PROBE
.DC TEMP -50 50 2.5
.END
```

When this netlist is run, the entire circuit is calculated at each temperature. Since the only component with a temperature coefficient is the RTD, this is the only component that will change with temperature.

6.1.3. 60-Hz Reject Filter

This is a simple unity-gain circuit that is used to eliminate 60-Hz noise from a circuit. The circuit is shown in Figure 6-4. The netlist for this filter is

```
*60 Hz REJECT FILTER DEMO
VCC 1 0 15
VEE 2 0 -15
VAC 3 0 AC 1
R1 5 7 270K
R2 7 8 270K
R3 6 9 150K
R4 3 5 1K
C1 5 6 .01U
C2 6 8 .01U
C3 7 0 .02U
*C4 4 5 100U
RL 9 0 10K
XAMP 9 8 1 2 9 UA741
.LIB
.AC DEC 100 1 1000
.PROBE
.END
```

The output of the filter versus frequency is shown in Figure 6-5.

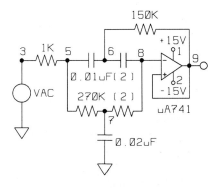

Figure 6-4

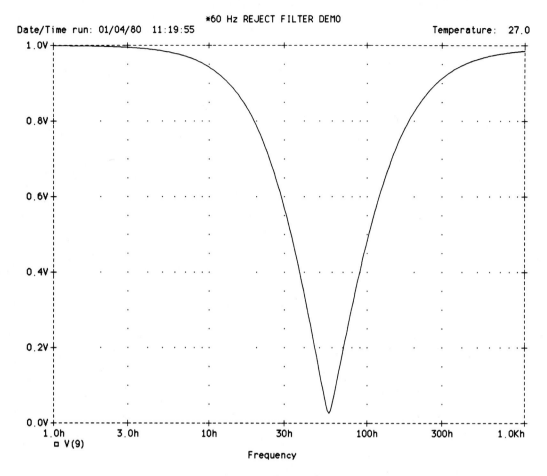

Figure 6-5

6.1.4. Op Amp Integrator

The active integrator is a low-pass filter. It will also change a positive going sine wave into a cosine wave, and it will change a square wave into a triangle wave. The output of the integrator is dependent on the time that an input signal is present. It will saturate eventually even if the input is very small but remains applied for a long time. An integrating circuit that is used to change a 1-kHz square wave into a triangle wave is analyzed next. The circuit is shown in Figure 6-6. Note that a pulse waveform could have been used for the input square wave also. The netlist for the circuit is

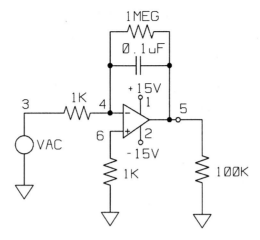

Figure 6-6

```
*OPAMP INTEGRATOR
VCC 1 0 15
VEE 2 0 -15
VAC 3 0 PWL(0,-1 1U,1 499U,1 500U,-1 .001,-1 1.001M,1
+      1.499M,1 1.5M,- 1 2M,-1)
R1 3 4 1K
C1 4 5 .1U
RX 5 4 1MEG
 R2 6 0 1K
RY 5 0 100K
XAMP 6 4 1 2 5 UA741
.TRAN 10U 2M UIC
.IC V(5) = 0
.PROBE
.LIB
.END
```

The output of this analysis is shown in Figure 6-7.

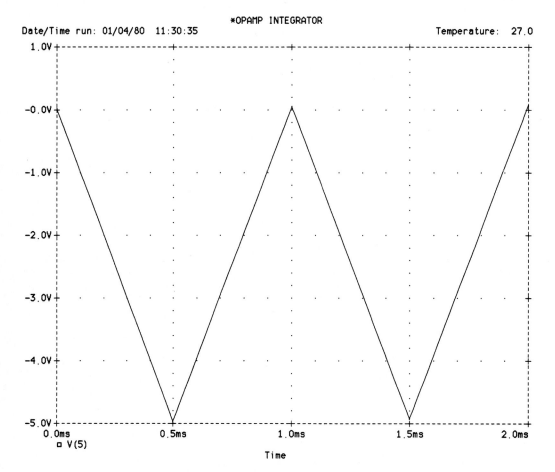

Figure 6-7

6.1.5. Op Amp Differentiator

The active differentiator is the opposite of the integrator in that its output depends on the rate of change of the input signal. Thus very slowly changing inputs produce little output, but inputs such as a square wave will produce a train of narrow positive- and negative-going output pulses. A triangle wave will produce a square wave since it is changing at a constant rate.

The analysis of a differentiator follows. The circuit of the differentiator is shown in Figure 6-8. The netlist for the differentiator is

```
*DIFFERENTIATOR CIRCUIT
VCC 1 0 15
VEE 2 0 -15
```

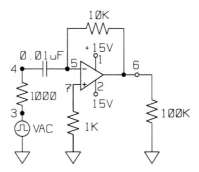

Figure 6-8

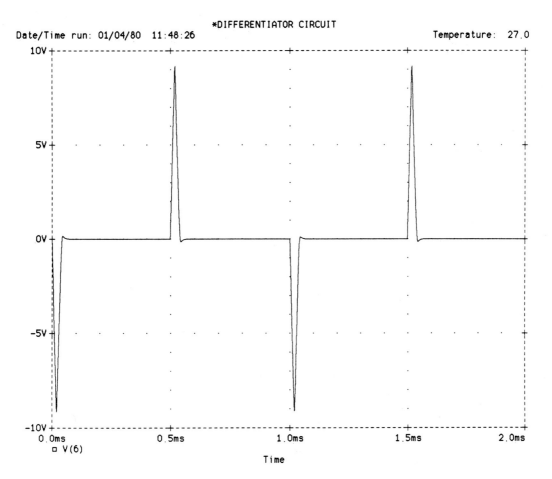

Figure 6-9

```
VAC 3 0 PULSE(-1 1 0 1U 1U 498U 1M)
R1 3 4 1000
R2 6 5 10K
R3 7 0 1K
R4 6 0 100k
C1 4 5 .01U
XAMP 7 5 1 2 6 UA741
.LIB
.TRAN .01U 2M
.PROBE
.END
```

The output of this analysis is shown in Figure 6-9.

6.1.6. Schmitt Trigger

The Schmitt trigger uses positive feedback to cause an input signal to drive the op amp to saturation quickly in both positive and negative directions. An example of the use of the Schmitt trigger is to cause sine wave pulses to become rectangular for input to a digital circuit. This circuit has two levels of triggering. One is for the positive slope of the input signal and the other for the negative slope of the input signal. There is some difference in these two levels, and this is called the circuit *hysteresis*. Hysteresis is usually the most important parameter of this circuit. The hysteresis can be viewed using PROBE by changing the x axis from time to voltage. The analysis that follows shows this effect, as well as the output signals of the circuit. The circuit of the Schmitt trigger is shown in Figure 6-10. The netlist for the Schmitt trigger is

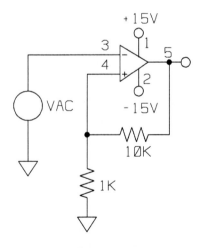

Figure 6-10

```
*SCHMITT TRIGGER
VCC 1 0 15
VEE 2 0 -15
VAC 3 0 SIN(0 5 1000)
R1 4 0 1000
R2 4 5 10000
XAMP 4 3 1 2 5 UA741
.TRAN .01M 2M
.PROBE
.LIB
.END
```

The output of this circuit is shown in Figure 6-11 and the hysteresis in Figure 6-12.

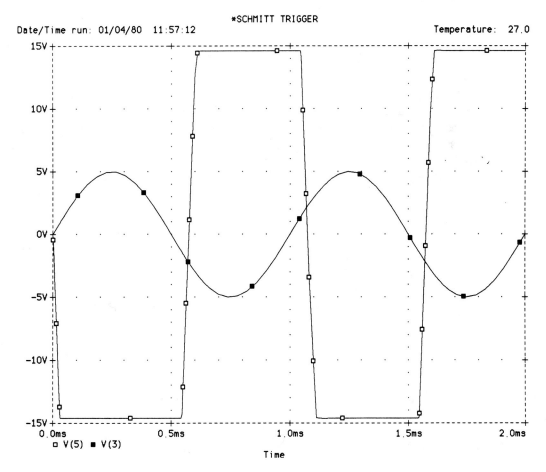

Figure 6-11

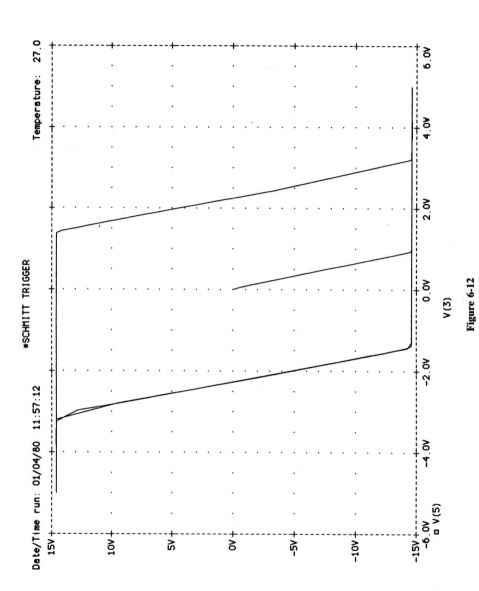

*SCHMITT TRIGGER

Date/Time run: 01/04/80 11:57:12

Temperature: 27.0

V(3)

□ V(5)

Figure 6-12

180

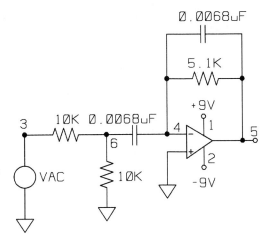

Figure 6-13

6.1.7. Bandpass Filter

This is a simple bandpass filter with a Q value of 0.505. The circuit of the filter is shown in Figure 6-13. The netlist for the filter is

```
*BANDPASS FILTER.
VAC 3 0 AC 1
R1 3 6 10K
R2 6 0 10K
R3 4 5 5.1K
C1 6 4 .0068U
C2 4 5 .0068U
XAMP1 0 4 1 2 5 UA741
VCC 1 0 9
VEE 2 0 - 9
.LIB
.AC DEC 10 10 1MEG
.PROBE
.END
```

The output of the analysis for both frequency and phase is shown in Figure 6-14.

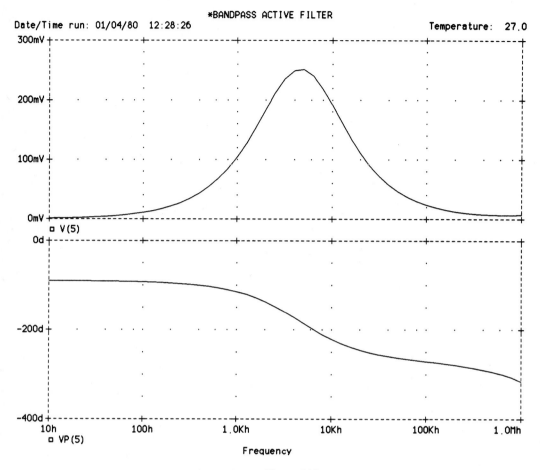

Figure 6-14

6.2. THE LM3900

Another amplifier circuit that is in wide use is the Norton, or current differencing amplifier. This is not a true op amp but is included in the op amp section of manufacturers' literature. The basic circuit for this amplifier is shown in Figure 6-15. The data sheet does not give enough information to duplicate this amplifier exactly. We can still make a model that performs close to the actual device. This is done using default transistors with some capacitance added to meet the frequency response needs of the amplifier.

The only device that we must be careful of is the diode that is used to bias the noninverting input. This diode should be a diode connected transistor of the same type used for the amplifier. This assures that the emitter–base voltages are most closely matched. This is important because all of the other voltages in the device will be in-

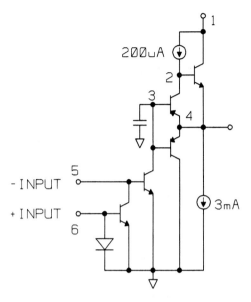

Figure 6-15

correct if this is not done. There is also a voltage supply for the VCCS devices. The netlist for the subcircuit is

```
*NORTON AMPLIFIER SUBCIRCUIT
.SUBCKT LM3900 6 5 1 4
I1 1 2 200U
I2 4 0 1.3M
Q1 1 2 4 QMOD1
Q2 4 3 2 QMOD2
Q3 0 3 4 QMOD2
Q4 3 5 0 QMOD1
Q5 5 6 0 QMOD1
Q6 6 6 0 QMOD1
C1 3 0 1P
.MODEL QMOD1 NPN(CJC=0.5P)
.MODEL QMOD2 PNP(CJC=0.5P)
.ENDS
```

To show the use of this amplifier, let us make a bandpass active filter with a center frequency of 1000 Hz and a Q value of 25. The circuit of the bandpass filter is shown in Figure 6-16. The netlist for the bandpass filter is

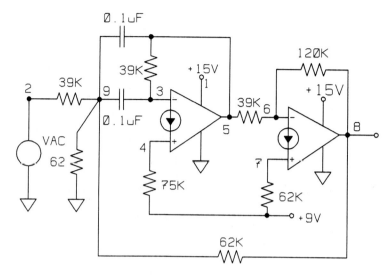

Figure 6-16

```
*NORTON AMPLIFIER BANDPASS FILTER
VCC 1 0 15
VAC 2 0 AC 1
R1 2 9 39K
R2 9 0 62
R3 5 3 39K
R4 5 6 39K
R5 8 6 120K
R6 4 1 75K
R7 7 1 62K
R8 9 8 62K
C1 9 5 .1U
C2 9 3 .1U
XAMP1 4 3 1 5 LM3900
XAMP2 7 6 1 8 LM3900
.LIB
.AC DEC 250 300 3K
.PROBE
.OP
.END
```

The dc and ac output of the bandpass filter is shown in Figure 6-17 and Analysis 6-2.

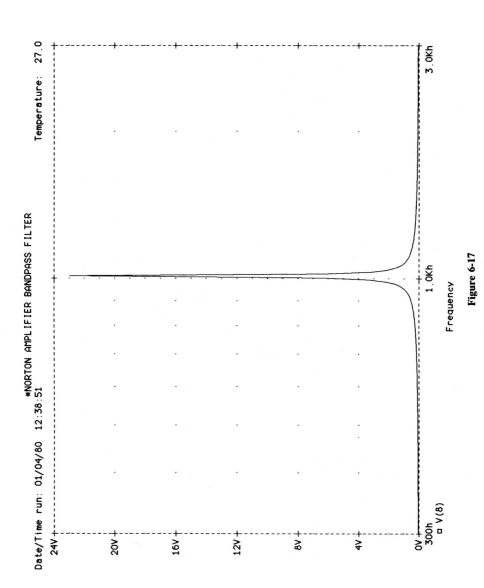

Figure 6-17

```
******* 01/04/80 ******* Evaluation PSpice (January 1990) ******* 12:38:51 *******
*NORTON AMPLIFIER BANDPASS FILTER
****      CIRCUIT DESCRIPTION
*************************************************************************
VCC 1 0 15
VAC 2 0 AC 1
R1 2 9 39K
R2 9 0 62
R3 5 3 39K
R4 5 6 39K
R5 8 6 120K
R6 4 1 75K
R7 7 1 62K
R8 9 8 62K
C1 9 5 .1U
C2 9 3 .1U
XAMP1 4 3 1 5 LM3900
XAMP2 7 6 1 8 LM3900
.LIB SUB.LIB
.AC DEC 250 300 3K
.PROBE
.OP
.END
```

```
******* 01/04/80 ******* Evaluation PSpice (January 1990) ******* 12:38:51 *******
*NORTON AMPLIFIER BANDPASS FILTER
****      BJT MODEL PARAMETERS
*************************************************************************
```

	XAMP1.QMOD1 NPN	XAMP1.QMOD2 PNP	XAMP2.QMOD1 NPN	XAMP2.QMOD2 PNP
IS	100.000000E-18	100.000000E-18	100.000000E-18	100.000000E-18
BF	100	100	100	100
NF	1	1	1	1
BR	1	1	1	1
NR	1	1	1	1
CJC	500.000000E-15	500.000000E-15	500.000000E-15	500.000000E-15

```
******* 01/04/80 ******* Evaluation PSpice (January 1990) ******* 12:38:51 *******
*NORTON AMPLIFIER BANDPASS FILTER
****      SMALL SIGNAL BIAS SOLUTION      TEMPERATURE =    27.000 DEG C
*************************************************************************
```

NODE	VOLTAGE	NODE	VOLTAGE	NODE	VOLTAGE	NODE	VOLTAGE
(1)	15.0000	(2)	0.0000	(3)	.6112	(4)	.7308
(5)	7.8865	(6)	.6116	(7)	.7357	(8)	5.2966
(9)	.0053	(XAMP1.2)	8.6707			(XAMP1.3)	7.9404
(XAMP2.2)	6.0761			(XAMP2.3)	5.3454		

```
    VOLTAGE SOURCE CURRENTS
    NAME          CURRENT

    VCC           -3.521E-03
    VAC           1.355E-07

    TOTAL POWER DISSIPATION    5.28E-02  WATTS
           JOB CONCLUDED
           TOTAL JOB TIME            75.85
```

Analysis 6-2

SUMMARY

Op amps in PSpice are not models but subcircuits in which models may be used. When specifying an op amp in a circuit, an X call for a subcircuit must be used. A complete op amp subcircuit is composed of eight active devices and uses about 18 nodes.

Op amps can be used with ideal op amp subcircuits to cut the node count in a circuit. Op amp subcircuits can be used in amplifiers, and oscillators and will perform substantially the same as an actual device in a circuit.

SELF-EVALUATION

1. For the notch filter shown in Figure P6-1, plot the frequency response as the capacitor is changed from 100 pF to 500 pF in 100-pF increments. For the circuit R1 = R2, R4 = R5 = R2/2, and $f = 1/2\pi\sqrt{C1C2(R5)^2}$. The supply voltages are +15 and −15 V.

2. The circuit shown in Figure P6-2 is a window comparator. Using a dc sweep show the output of this amplifier as the input voltage is swept between −10 and 10 V in 0.1-V increments. The op amp supply voltages are +15 and −15 V.

3. The circuit shown in Figure P6-3 is an precision rectifier. Using a 1-kHz sine wave, show that the output of this circuit is the negative peaks of sine wave. Analyze over a 10-ms period. The op amp supply voltages are +15 and −15 V.

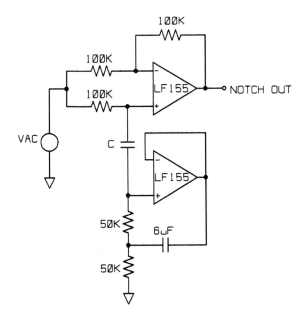

Figure P6-1

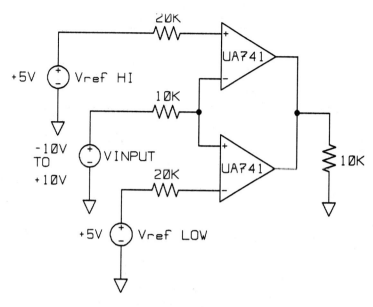

Figure P6-2

4. The circuit shown in Figure P6-4 is a monostable form of circuit. Show that this circuit will divide an input pulse train by 4. Use a pulse waveform for the input.

5. The circuit shown in Figure P6-5 is an active bandpass filter. Show both the frequency and phase of this amplifier for the range 1 Hz to 100 kHz. Use a decade form of sweep with 25 points per decade.

6. The circuit shown in Figure P6-6 is an equal-component Sallen–Key filter. Design the filter to have a cutoff of 5 kHz. The ideal gain for maximum flatness in the passband is 1.586. Design the filter to have three gains: 1, 1.586, and 2.5. Use a decade type of sweep from 1

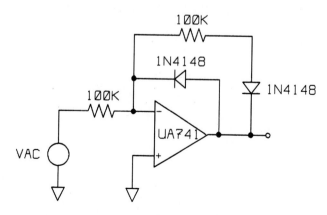

Figure P6-3

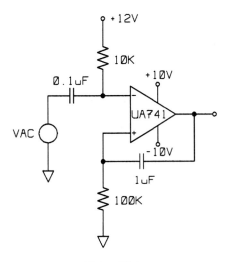

Figure P6-4

Hz to 100 kHz. Plot the frequency and phase at the output of the filter. Show that for the ideal gain of the amplifier, the filter has a rolloff of 40 dB per decade.

7. For the circuit shown in Problem 6, use an ideal amplifier and show that the responses are similar for the same conditions as in Problem 6.

8. For the audio amplifier in Section 6.1.1, set this circuit up to operate from $\pm$ 9-V supplies. Show that the output is about twice that of the single supply circuit.

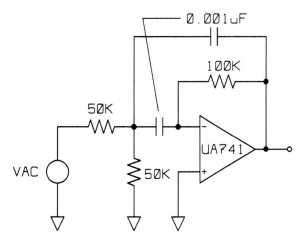

Figure P6-5

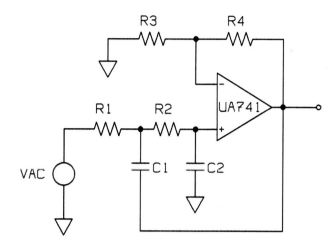

Figure P6-6

9. The circuit shown in Figure P6-9 is a differential amplifier with a common-mode signal. The manufacturer specifies that the CMRR is 70 dB minimum. Show that the CMRR of the amplifier is within that specified by the manufacturer.

10. The circuit shown in Figure P6-10 is a zener diode limited amplifier. Plot the transfer curve of this amplifier as the input voltage is swept between ± 2 V.

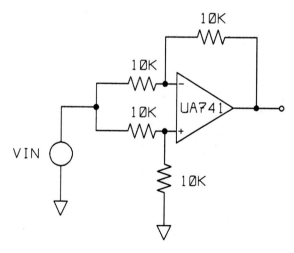

Figure P6-9

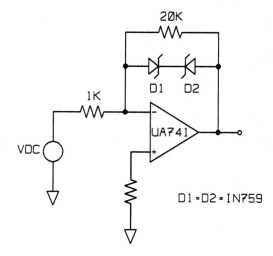

Figure P6-10

7

Oscillators

OBJECTIVES

1. To learn to use PSpice to analyze *RC* and *LC* oscillator circuits.
2. To learn methods to assure that the oscillators will start and free-run.

INTRODUCTION

Oscillators are in use in everything that we work with on a daily basis. From radios to computers—all use oscillators as part of their basic operating system. In this chapter we investigate several types of oscillator, from simple *RC* relaxation types that produce complex waveforms to audio and RF types. We will use both discrete components and integrated circuits for the oscillators.

7.1. *RC* RELAXATION OSCILLATORS

These oscillators produce complex waveforms. We include in these waveforms, the square wave and pulse, sawtooth, and triangle waves. The oscillators we design use both transistors and op amps. The design theory for these oscillators is well known and will not be presented here.

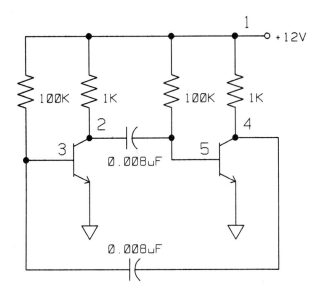

Figure 7-1

7.1.1. Square and Rectangular Wave (Pulse) Generators

The simplest of the rectangular wave generators is the astable multivibrator using two transistors. To assure that this or any other oscillating circuit will work, you must set some initial conditions in the netlist. The reason for this is that PSpice considers transistors and other components to be perfect devices. This means that there is no difference between the transistors, or components that are specified. Thus there is no difference in the time it takes either transistor and their *RC* components to reach either cutoff or saturation. Because of this, both collectors can be either high or low. We will use specific initial conditions to set the collectors to the desired state before analysis.

To set the initial conditions you can use either the .IC command or the .NODE-SET command. The circuit diagram of a two-transistor multivibrator is shown in Figure 7-1. The frequency of a symmetrical astable multivibrator is

$$f = \frac{0.7}{R_B C}$$

Thus the frequency of this oscillator is about 875 Hz. The netlist for the multivibrator is

```
*TWO TRANSISTOR ASTABLE MULTIVIBRATOR
VDC 1 0 12
R1 1 2 1K
R2 1 4 1K
RB2 1 5 100K
RB3 1 3 100K
C1 2 5 .008U
```

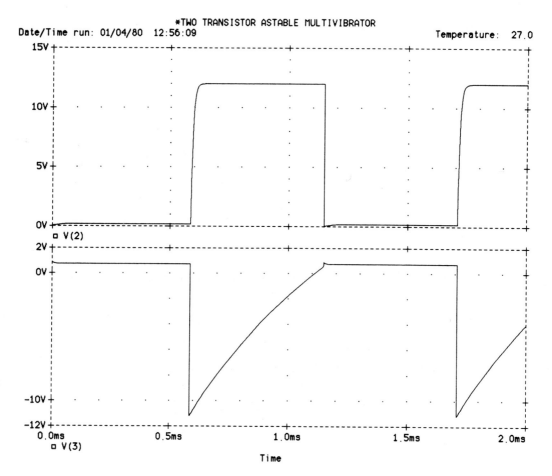

Figure 7-2

```
C2 4 3 .008U
Q1 2 3 0 Q2N3904
Q2 4 5 0 Q2N3904
.TRAN .01M 2M UIC
.IC V(4)=0 V(2)=12
.PROBE
.LIB
.END
```

The result of the analysis is shown in Figure 7-2.

A circuit that does the same job as the two-transistor astable multivibrator is the single op amp square wave generator. This is the next circuit to be analyzed. Note that the two resistors at the noninverting input have an effect on frequency. If you set the noninverting input to a small value of voltage, the oscillator frequency will be higher than if the voltage at the input is a larger value. The frequency of oscillation of this oscillator is

$$f = \frac{1}{2RC \ln\left(\dfrac{1+\beta}{1-\beta}\right)}$$

where $\beta = $ R3/(R2 + R3). For this oscillator this is about 1400 Hz. The circuit diagram for this analysis is shown in Figure 7-3. The netlist for this oscillator is

```
*INTEGRATED CIRCUIT SQUARE WAVE GENERATOR
VCC 1 0 15
VEE 2 0 -15
R1 5 3 100K
C1 3 0 .0033U
R2 5 4 100K
R3 4 0 100K
XAMP 4 3 1 2 5 LM149
.TRAN .01M 2M UIC
.IC V(3)=0
.LIB
.PROBE
.END
```

The output of this analysis is shown in Figure 7-4.

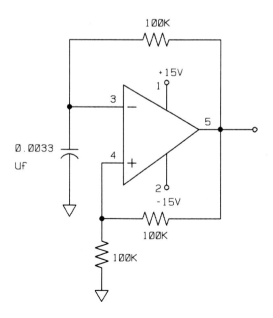

Figure 7-3

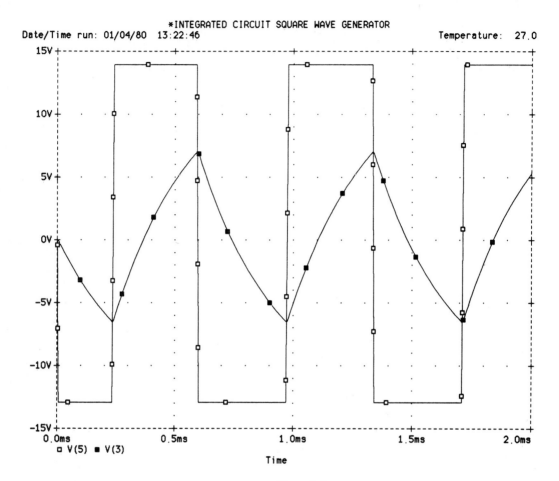

Figure 7-4

7.1.2. Pulse Generator

This circuit produces unsymmetrical output pulses. It is similar to the square wave generator but uses two paths for the charge and discharge of the timing capacitor. Diodes are used to determine the charge and discharge path of the timing capacitor. The circuit diagram to be used for this analysis is shown in Figure 7-5. The netlist for this circuit is

```
*RECTANGULAR PULSE GENERATOR
VCC 1 0 10
VEE 2 0 -10
R1 3 7 5K
```

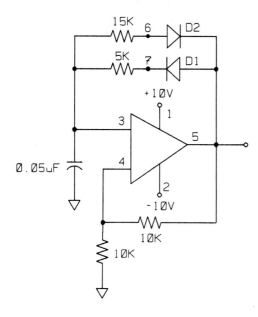

Figure 7-5

```
R2  3 6 15K
R3  4 0 10K
R4  4 5 10K
C1  3 0 .05U
D1  5 7 D1N914
D2  6 5 D1N914
XAMP 4 3 1 2 5 LM149
.LIB
.TRAN .01M 2M UIC
.IC V(5)=0
.PROBE
.END
```

The output of this analysis is shown in Figure 7-6.

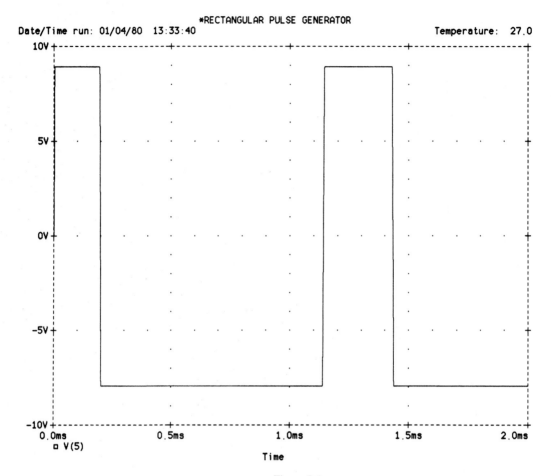

Figure 7-6

7.1.3. Triangle Wave Generator

The triangle wave generator using op amps is the combination of a square wave generator and an integrator. Actually, we have both a square and a triangle wave output available to us. This gives us a single-frequency function generator. It is assumed that you have the ideal model of the LF155 in the library file. The output voltage change of this device is

$$\Delta e_o = \frac{(-e_{in})T/2}{R_1 C_1}$$

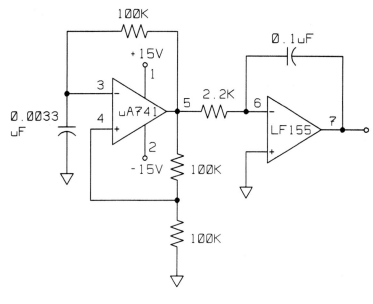

Figure 7-7

The circuit diagram of the triangle wave generator is shown in Figure 7-7. The netlist for this generator is

```
*INTEGRATED CIRCUIT TRIANGLE WAVE GENERATOR
VCC 1 0 15
VEE 2 0 -15
R1 5 3 100K
C1 3 0 .0033U
R2 5 4 100K
R3 4 0 100K
R4 5 6 2.2K
C2 6 7 .1U
XAMP 4 3 1 2 5 UA741
XAMP1 0 6 7 LF155
.TRAN 10U 2M UIC
.IC V(5)=0 V(3)=0
.PROBE
.LIB
.END
```

The output of this analysis is shown in Figure 7-8.

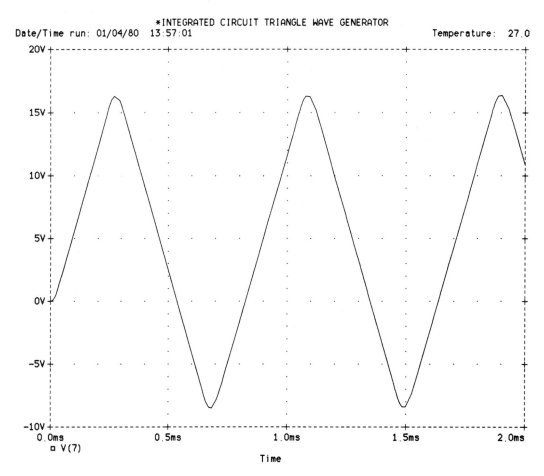

Figure 7-8

7.1.4. Simple VCO Using an Ideal Op Amp and a Macromodel Op Amp

This circuit uses an ideal op amp and a macromodel op amp for two reasons. First, the student version of PSpice will not handle this circuit because there are too many devices and nodes. Second, it points up that ideal op amps can be used for devices other than amplifiers and filters. This saves us some nodes and calculation time since the calculation time increases roughly as the square of the number of nodes. The change in frequency is accomplished by supplying the input with a piecewise linear waveform that rises from 0 to 20 V in 80 ms. As the input voltage is rising, the oscillator is also generating its own output. This produces an output that changes frequency as the voltage at its input changes. Once again it is assumed that the ideal op amp LF155 is in

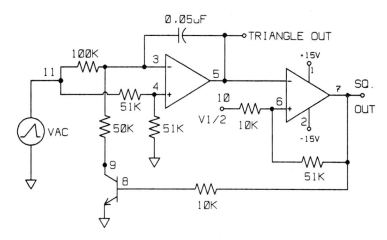

Figure 7-9

the library file. The circuit diagram of the VCO is shown in Figure 7-9. The netlist for the VCO is

```
*VCO
VCC 1 0 15
VEE 2 0 -15
VAC 11 0 PWL(0,0 80M,20)
V1/2 10 0 7.5
R1 11 3 100K
R2 3 9 50K
R3 11 4 51K
R4 4 0 51K
R5 10 6 10K
R6 6 7 51K
R7 7 8 10K
C1 3 5 .05U
XAMP 3 4 5 LF155
XAMP1 6 5 1 2 7 LM149
Q1 9 8 0 QMOD
.MODEL QMOD NPN
.OPTIONS RELTOL=0.01
.TRAN .1M 80M 20M UIC
.IC V(5)=0 V(3)=0
.PROBE
.LIB
.LIB SUB.LIB
.END
```

The output of the VCO analysis is shown in Figure 7-10.

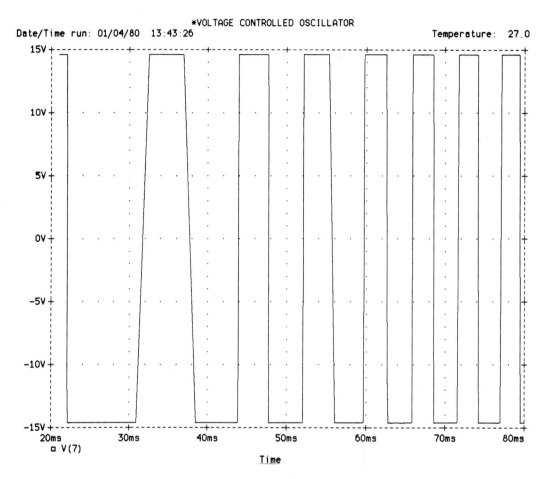

Figure 7-10

7.1.5. Op Amp Sawtooth Oscillator

The sawtooth oscillator is used in several electronic devices. Among these are oscillo-scopes, to sweep the electron beam across the CRT, and in the TV set for a similar application. The sawtooth is also used in spectrum analyzers for sweeping the local oscillator of the instrument. For this analysis, the fall time of the sawtooth wave is 10 times faster than the rise time of the sawtooth. The sawtooth oscillator circuit diagram is shown in Figure 7-11. The diodes in the circuit steer the charging currents for the capacitor. For a positive-going sawtooth, $R_3 < < R_4$. The netlist for the sawtooth generator is

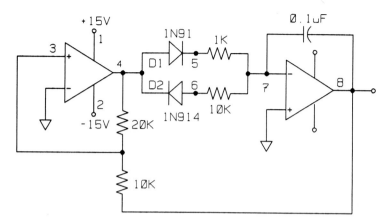

Figure 7-11

```
*SAWTOOTH GENERATOR
VCC 1 0 15
VEE 2 0 -15
R1 4 3 20K
R2 3 8 10K
R3 5 7 1K
R4 6 7 10K
C1 7 8 .1U
XAMP1 3 0 1 2 4 LM149
XAMP2 0 7 1 2 8 LM149
D1 4 5 D1N914
D2 6 4 D1N914
.LIB
.PROBE
.TRAN .01M 4M UIC
.IC V(8)=0 V(3)=0
.END
```

The output of this analysis is shown in Figure 7.12.

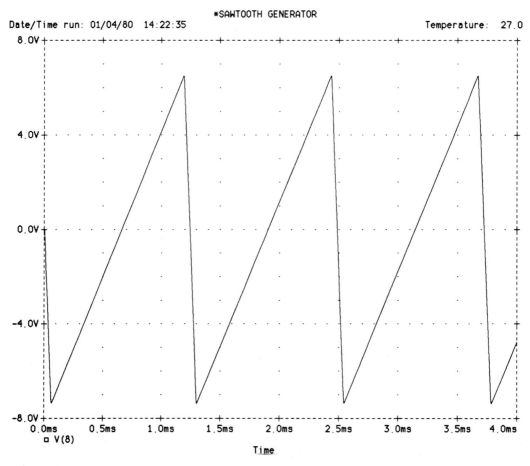

Figure 7-12

7.2. *RC* SINE WAVE OSCILLATORS

In this part of our examination of oscillators using PSpice, we will analyze three types of RC sine wave oscillator. We will analyze a phase-shift oscillator, a Wien bridge oscillator, and a quadrature oscillator. Once again, we must use either initial condition or nodeset command lines to assure that the oscillator will start. The design methods for these oscillators is well known and will not be presented here.

7.2.1. Phase-Shift Oscillator

This oscillator generates a frequency of about 4500 Hz. For the oscillator to run, a gain of about 29 is required in the op amp. The phase shift of each *RC* section is about 60°. This adds 180° phase shift from output to input. Thus the feedback is connected to the

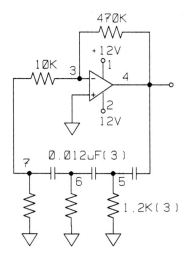

Figure 7-13

inverting input of the op amp. The output frequency of a three section phase-shift os-cillator is

$$f = \frac{1}{2\pi\sqrt{6}\ RC}$$

Figure 7-13 shows the circuit diagram of this oscillator. The netlist for this oscillator is

```
*PHASE-SHIFT OSCILLATOR
VCC 1 0 12
VEE 2 0 -12
R1 3 7 10K
R2 4 3 470K
R3 5 0 1.2K
R4 6 0 1.2K
R5 7 0 1.2K
C1 5 4 .012U
C2 6 5 .012U
C3 7 6 .012U
XAMP 0 3 1 2 4 UA741
.LIB
.PROBE
.TRAN .01M 1M 0 .01M UIC
.IC V(4)=-15
.END
```

The output of this analysis is shown in Figure 7.14.

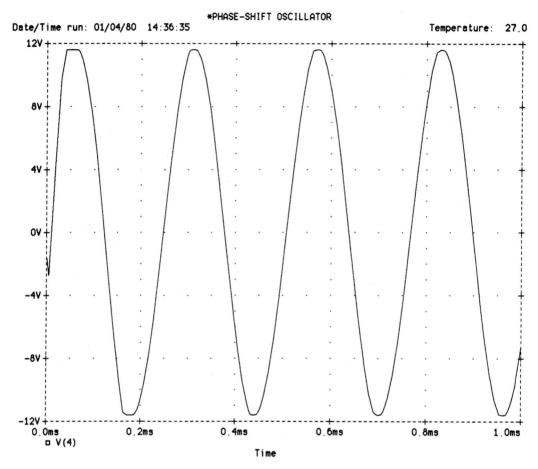

*PHASE-SHIFT OSCILLATOR
Date/Time run: 01/04/80 14:36:35   Temperature: 27.0

Figure 7-14

7.2.2. Wien Bridge Oscillator

The Wien bridge type of oscillator is one in which there are two RC filters. One is a high pass and the other a low pass. This produces a bandpass configuration with a specific center frequency. The form of this filter is similar to a Wheatstone bridge, with the op amp between the null points. Simple analysis shows the voltage output of the bandpass filter to be about one-third of the voltage input at the resonant frequency of the bridge. This requires that the amplifier have a gain of 3 to produce a loop gain of 1 in the circuit (Barkhausen's criteria). The output frequency of this oscillator is

$$f_{\text{RES}} = \frac{1}{2\pi RC}$$

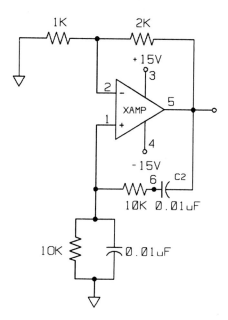

1K   2K

+15V

XAMP

-15V

C2

10K 0.01uF

10K   0.01uF

Figure 7-15

when the components of the *RC* network arms of the oscillator are equal values. The circuit diagram of the oscillator is shown in Figure 7-15. The netlist of the amplifier is

```
*WIEN BRIDGE SINE WAVE GENERATOR
V1  3  0  15
V2  4  0  -15
R1  0  2  1K
R2  2  5  2K
R3  1  0  10K
R4  1  6  10K
C1  1  0  .01U
C2  5  6  .01U
XBLOCK  1  2  3  4  5  UA741
.IC  V(5)=-15
.OP
.LIB
.TRAN  .05M  2.5M  0  .025M
.PROBE
.END
```

The output of this analysis is shown in Figure 7-16.

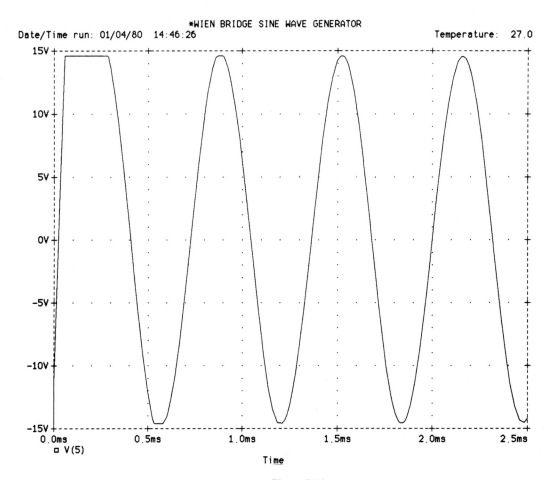

Figure 7-16

7.2.3. Quadrature Oscillator

This type of oscillator produces two sine wave outputs that are 90° out of phase with one another. Since one of the devices is used only for integration, we can take advantage of this and use the ideal op amp that we created earlier. Remember, the integrator only produces a phase shift of 90° in a sine wave. The diodes across the capacitor of the output amplifier are used to assure that the sine wave output does not clip. The circuit diagram of the oscillator is shown in Figure 7-17. The netlist of the oscillator is

```
QUADRATURE OSCILLATOR
VCC 1 0 15
VEE 2 0 -15
R1 3 0 20K
```

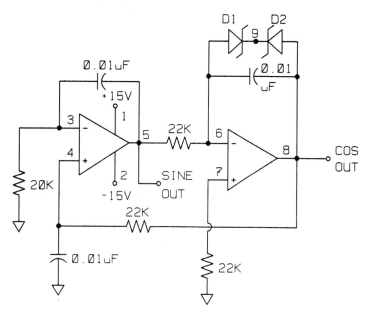

Figure 7-17

```
R2 5 6 22K
R3 8 4 22K
R4 7 0 22K
C1 4 0 .01U
C2 3 5 .01U
C3 6 8 .01U
D1 8 9 DMOD
D2 6 9 DMOD
XAMP1 4 3 1 2 5 UA741
XAMP2 7 6 8 LM155
.MODEL DMOD D(BV=10)
.LIB
.LIB SUB.LIB
.PROBE
.TRAN .01U 2.5M 0 .01M UIC
.IC V(5)=0 V(8)=15
.END
```

The output of this analysis is shown in Figure 7-18.

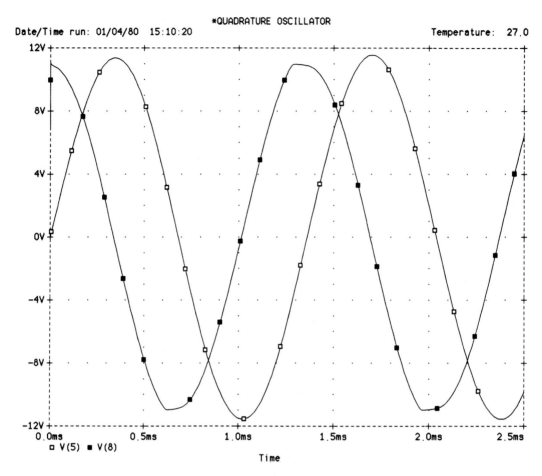

Figure 7-18

7.3. *LC* OSCILLATORS

In this part of the chapter we investigate three types of *LC* oscillator: the Armstrong, Hartley, and Colpitts oscillators. These oscillators are used in radio receivers, transmitters, and other equipment. Running the analysis of these types of oscillator is somewhat different in that there are two ways in which we can assure that the oscillator starts and runs. The first way is to use the UIC command to set certain nodes at zero, or the supply voltage. This can be tricky because you must select the nodes carefully. The second way is to insert a voltage generator in one of the leads of the oscillator and give a very short pulse to start the oscillation. The generator has no effect after the pulse because all voltage generators have zero internal impedance.

If you use the UIC method, the oscillator may also damp out if the initial conditions are chosen incorrectly. This can cause the entire analysis time to be occupied by transient conditions only. You may have to try several different initial conditions before you find the one that works.

    Since there is a transient at the start of oscillator operation, you can have PROBE not report the results of calculation until a specific amount of time has elapsed (see the form of the .TRAN command in Chapter 5). It is tempting to use the inductor and capacitor in a tuned circuit without considering the resistance of the coil. Remember that in PSpice the inductor is perfect. This means the inductor has zero resistance, giving the coil itself a *Q* of infinity. Always include a few ohms of resistance in series with the inductor to provide for skin effect and real resistance.

7.3.1. Armstrong Oscillator

This type of oscillator uses inductive feedback in the form of a transformer winding to the base of the transistor. This produces the proper phase of feedback for oscillation. The connections of the transformer winding must be properly arranged for the correct polarity of feedback to occur. The UIC method of starting the oscillator is used. The output frequency of this oscillator is

$$f_o = \frac{1}{2\pi\sqrt{L_1 C_1}}$$

The output frequency of this oscillator is approximately 160 kHz. The circuit diagram of the oscillator is shown in Figure 7-19. The netlist of the oscillator is

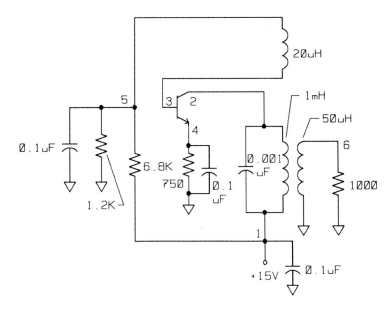

Figure 7-19

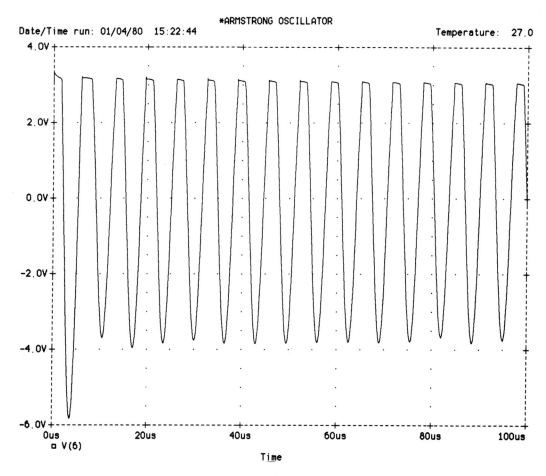

Figure 7-20

```
*ARMSTRONG OSCILLATOR
VCC 1 0 15
R1 1 5 6.8K
R2 5 0 1.2K
R3 4 0 750
R4 6 0 1K
C1 2 0 .001U
C2 4 0 .1U
C3 5 0 .1U
C4 1 0 .1U
L1 1 2 .001
L2 3 5 20U
L3 6 0 50U
```

```
K123 L1 L2 L3 0.9999
Q1 2 3 4 Q2N3904
.LIB
.OPTIONS RELTOL=0.01
.PROBE
.TRAN/OP 1U 100U 0 .15U UIC
.IC V(2)=0
.END
```

The output of this analysis is shown in Figure 7-20.

7.3.2. Hartley Oscillator

This type of oscillator also uses inductive feedback, but from a tapped coil. The output frequency of this oscillator is given by a similar equation to the one for the Armstrong oscillator. Only the inductance must be modified to include both of the components of the inductance and the mutual inductance, L_M ($L_1 + L_2 + 2L_M$). The pulse method of starting is used for this oscillator. The name of the starting pulse generator is VSTART. The output frequency of this oscillator is approximately 2.25 MHz. The circuit diagram of this oscillator is shown in Figure 7-21. The netlist for this oscillator is

```
*HARTLEY OSCILLATOR DEMO FILE
VCC 1 0 12
R1 1 3 10K
R2 3 0 2K
R3 4 0 500
R4 6 0 500
C1 1 0 0.001U
C2 2 0 100P
C3 4 5 .001U
```

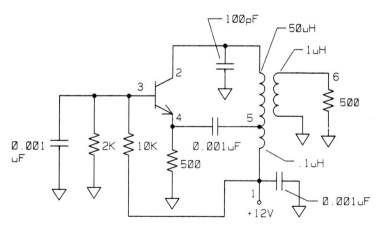

Figure 7-21

```
C4 3 0 .001U
L1 1 5 .1U
L2 5 2 50U
L3 6 0 1U
K12 L1 L2 L3 .9999
Q1 2 3 4 Q2N3904
VSTART A 0 PWL(0,0 1N,12 5N,12 6N,0) ;starting pulse
.LIB
.PROBE
.TRAN .1U 5U 0 .05U
.END
```

The output of this analysis is shown in Figure 7-22. Note that the *x* and *y* axes have
been rescaled. This oscillator shows clipping of the output waveform. This is probably
due to excessive feedback. You should try to improve the output of this oscillator.

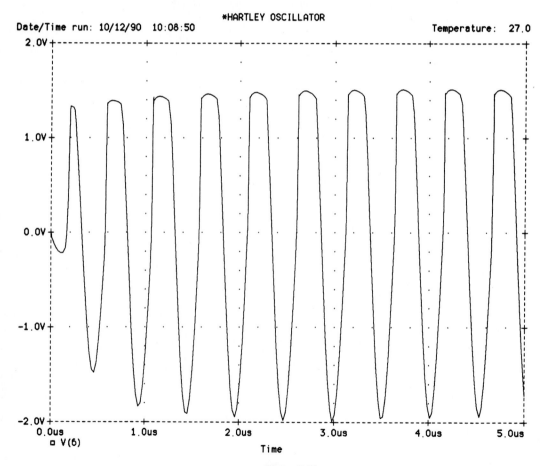

Figure 7-22

7.3.3 Colpitts Oscillator

This oscillator uses capacitive feedback in the form of a pair of capacitors in the tuning circuit. The capacitors are arranged as a voltage divider to provide the necessary feedback to the input of the amplifying system. The output frequency of this oscillator is

$$f_o = \frac{1}{2\pi\sqrt{L_1 C_T}}$$

where $C_T = (C_1 C_2)/(C_1 + C_2)$. The output frequency of this oscillator is about 9.6 MHz. The circuit diagram of the Colpitts oscillator is shown in Figure 7-23. The netlist of the oscillator is

```
*COLPITTS OSCILLATOR DEMO FILE
V1  1  0  12
R1  1  3  10K
R2  3  0  2K
R3  4  0  1000
R4  4  5  82
C1  2  5  150P
C2  5  0  1500P
C3  2  0  30P
C4  3  0  .01U
C5  1  0  .01U
L1  2  1  2U
L2  6  0  .2U
RL  6  0  600
```

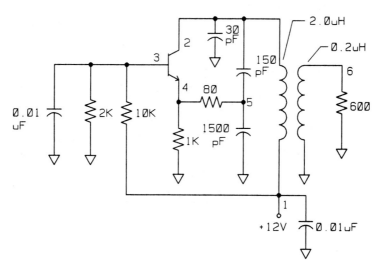

Figure 7-23

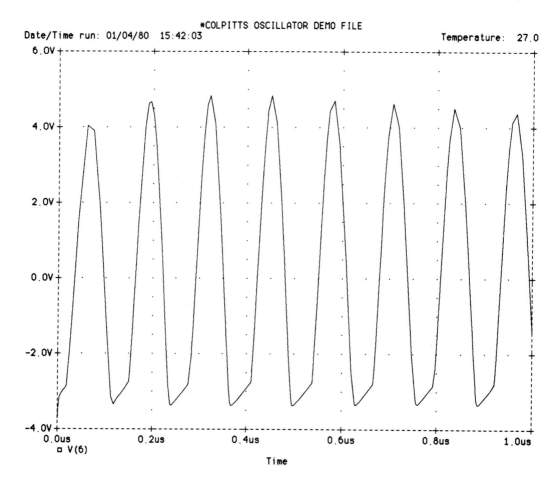

Figure 7-24

```
K12 L1 L2 .9999
Q1  2  3  4  Q2N2222A
.LIB
.TRAN .05U 1U
.IC V(4)=0
.PROBE
.END
```

The output of this analysis is shown in Figure 7-24.

SUMMARY

PSpice can work with circuits that generate their own signals. In this type of circuit it is necessary to provide the necessary starting conditions for the oscillator. There are two methods of starting an oscillator. The user can specify initial conditions that change after the bias conditions have been calculated, causing the oscillator to start, or a voltage generator with a short pulse can be used to supply the starting pulse for the oscillator. Both *RC* and *LC* types of oscillator can be run using PSpice. In both cases, starting conditions must be supplied.

SELF-EVALUATION

1. Figure P7-1 shows a simple function generator made of two μA741 op amps. Analyze this circuit and show the outputs at both the comparator and integrator.
2. Figure P7-2 is a form of Armstrong oscillator. Analyze this circuit and show the output at the collector.
3. Figure P7-3 shows a form of sine wave oscillator called a bridged-T oscillator. Analyze this circuit and show the sine wave at the output of the μA741.
4. Figure P7-4 is a form of gated sine wave oscillator. Analyze this circuit for a total time of 30 ms.
5. Figure P7-5 is a JFET Hartley oscillator. Analyze this circuit and show the output across the 1.5-kΩ load resistor.

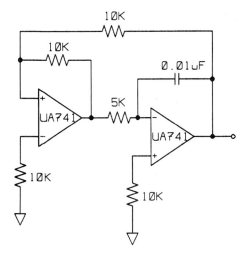

Figure P7-1

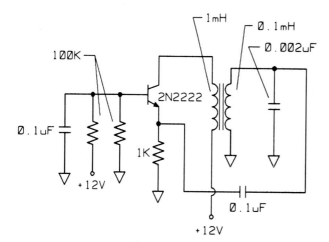

Figure P7-2

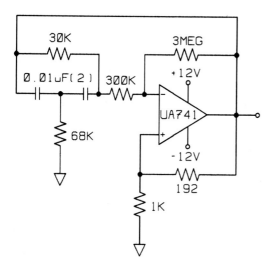

Figure P7-3

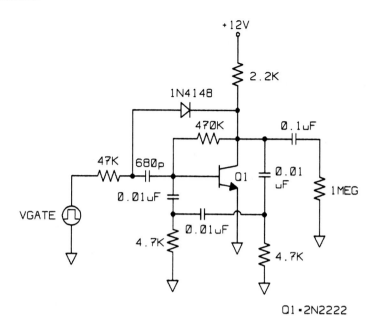

Figure P7-4

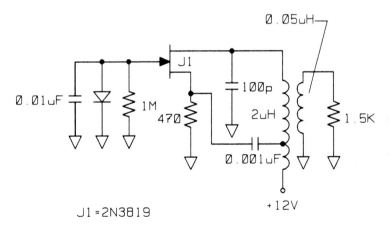

Figure P7-5

8

Feedback Control Analysis

OBJECTIVES

1. To learn to use PSpice in simple feedback control systems.
2. To use PSpice to produce the Nichols chart output and time-domain analysis.

INTRODUCTION

Feedback control systems are part of all areas of technology, from electrical and mechanical engineering to sociological and business systems. The techniques of analysis using the classical transfer functions and state-variable techniques are the most used forms of analysis. PSpice can perform many of the simulations needed in feedback analysis.

8.1. FUNDAMENTAL FEEDBACK SYSTEM

There are two basic feedback control system block diagrams: the unity-gain system with feedback equal to 1, and the system with feedback less than 1. Both of these systems are shown in Figure 8-1. The transfer function for each of these systems is

$$\frac{C(s)}{R(s)} = \frac{G(s)}{1+G(s)} \qquad (A_v = 1 \text{ system})$$

$$\frac{C(s)}{R(s)} = \frac{G(s)}{1+G(s)H(s)} \qquad (A_v > 1 \text{ system})$$

In some cases frequency response analysis is used to analyze these systems. For frequency response analysis, the Laplacian operator s can be replaced with $j\omega$, and the response calculated. As an example, the case of a simple RC integrating network response is given by

$$\frac{v_o}{v_i} = \frac{1}{1+sRC}$$

$$= \frac{1}{1+j\omega RC}$$

The response of this type of circuit is given by a magnitude (in dB) and a phase angle. When $j\omega RC$ is equal to 1, the output of the circuit falls to -3 dB, and the phase angle of the circuit is $-45°$. At $10f_{-3\text{dB}}$ the output of the circuit is 0.1, or -20 dB, of the output at low frequencies. You have seen these curves before, so they will not be shown here. Even if we add an amplifier to the system, we would not gain any more information about the system. The system is stable for all gains.

It is more informative to analyze a system that has a second-order response. Such a system is a simple second-order filter of the Butterworth type. The filter to be analyzed is the equal-component Sallen–Key low-pass filter. This filter has an optimum gain of 1.586 for maximal flatness of response. If the gain is increased above this value, the filter has a peaked response and may oscillate. We do not usually think of this

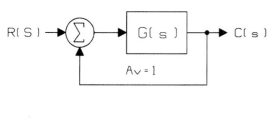

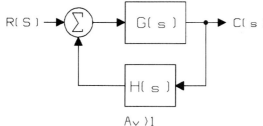

Figure 8-1

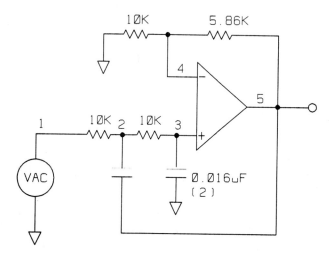

Figure 8-2

circuit as a feedback control system, but it is. It is instructive to analyze and then look at the Bode plots. PROBE can show both the real and imaginary components of a voltage, as well as the phase angle of the voltage. We can make the x axis of PROBE the phase of the output voltage, and the y axis the decibel magnitude of the output voltage. This makes what is called a Nichols chart, after its designer N. B. Nichols. The circuit we shall use is shown in Figure 8-2. The netlist for the analysis is

```
*SALLEN-KEY FILTER ANALYSIS
VAC 1 0 AC 1
R1 1 2 10K
R2 2 3 10K
R3 4 5 5.86K
R4 4 0 10K
C1 2 5 .016U
C2 3 0 .016U
XAMP 3 4 5 LF155
.AC DEC 10 10 100K
.PROBE
.LIB SUB.LIB
.END
*SALLEN-KEY FILTER ANALYSIS
VAC 1 0 AC 1
R1 1 2 10K
R2 2 3 10K
R3 4 5 15K
R4 4 0 10K
C1 2 5 0.16U
C2 3 0 .016U
XAMP 3 4 5 LF155
```

```
.AC DEC 10 10 100K
.PROBE
.LIB SUB.LIB
.END
```

Note the change in gain between the two analyses. The second analysis has the filter with a gain of 2.5, which gives a peak of about 12 dB. The gain and phase graphs of the filter are shown in Figures 8-3 and 8-4.

We could lay a straightedge on these plots and find the gain and phase margins of the systems. It is easier to use the Nichols chart for this purpose. All that is necessary is to rescale the axes of PROBE. We can have the *x* axis become the phase of the output voltage at node 5. Then the *y* axis becomes the decibel output of the filter at node 5 also. You may scale the axes further to facilitate the Nichols chart that you are

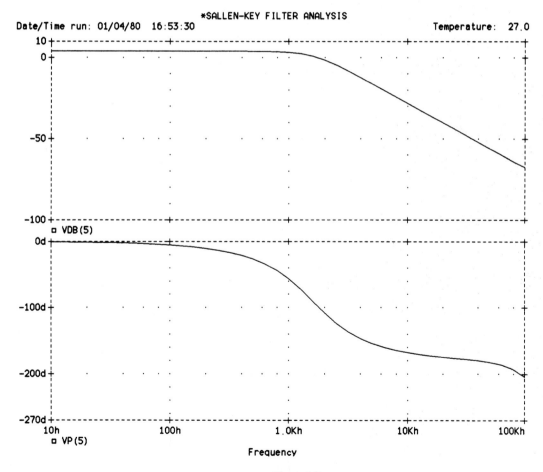

Figure 8-3

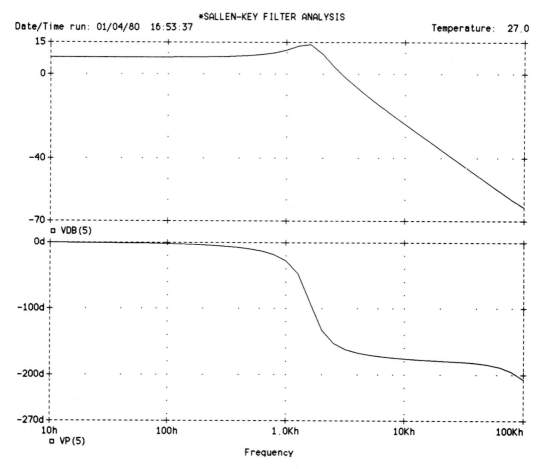

Figure 8-4

using. The Nichols charts for the active filter for both values of gain are shown in Figures 8-5 and 8-6. Looking at these charts clearly indicates that the circuit with a gain of 2.5 is becoming unstable, and the circuit with optimum gain is stable. Usually, if the system has a gain at its peak of +3 dB or less, the system is considered to be stable.

8.2. SECOND-ORDER SYSTEM ANALYSIS IN THE TIME DOMAIN

Another of the classical methods of analyzing feedback control systems is by simulating the system in the time domain. This is done using a step input or impulse. Any system whose differential equation has constant coefficients can be analyzed by this method. Second-order systems can have one of three distinct types of response:

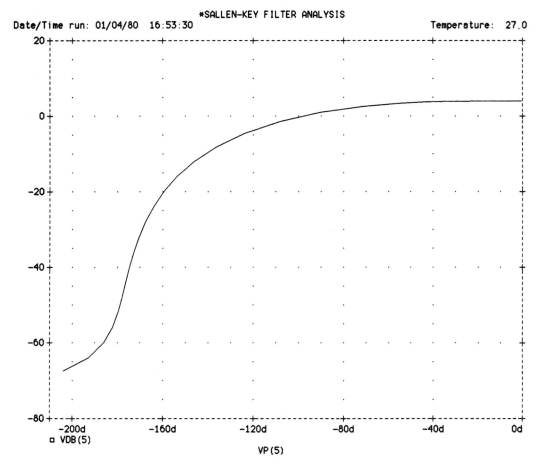

Figure 8-5

1. An underdamped response
2. A critically damped response
3. An overdamped response

The underdamped condition is of special importance since it is used to produce a system with a given amount of overshoot and a specified risetime and settling time. This speeds the response of the system to an input. With PSpice, we can use the ideal op amps that we have made to simulate the performance of such a system. The system may be mechanical, electrical, or any other type of system.

The first system we shall analyze is a torsional mechanical spring system. We will input an angular displacement and determine the time required for the system to come to rest. Simulation of this system is most easily done using analog simulation devices. In this case we use integration amplifiers to analyze how this system operates.

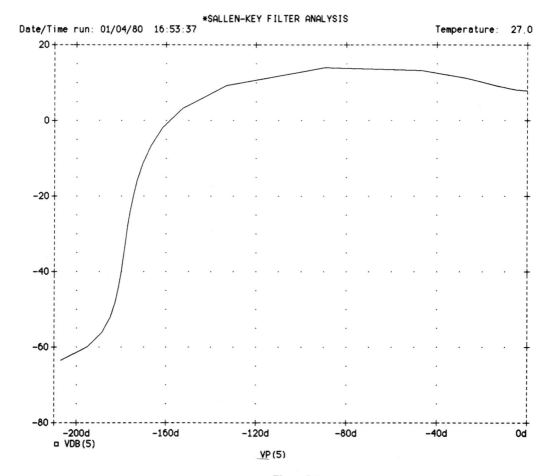

Figure 8-6

8.2.1. Torsion System Analysis

The mass is a solid cylinder weighing 16 lb and having a radius of 12 in. The inertia of the cylinder is

$$J = 0.5Mr^2 = 0.5\left(\frac{16}{32}\right)1^2 = 0.25 \text{ slug-ft}^2$$

The constant for the spring is 2 lb-ft/rad, and the damping is 0.4 lb-ft/(rad/s). From these constants we determine the differential equation to be

$$0.25 \frac{d^2\theta_o}{dt^2} + 0.4 \frac{d\theta_o}{dt} + 2\theta_o = 2\theta_i$$

Clearing the second derivative and isolating it, we have

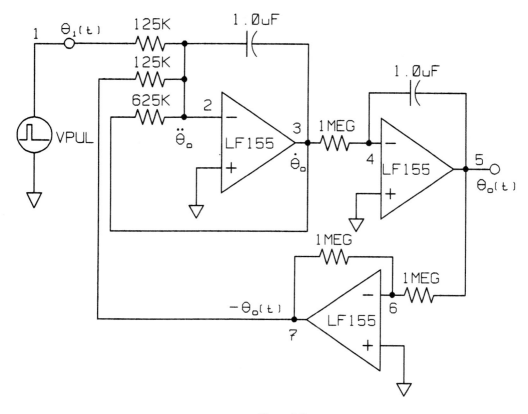

Figure 8-7

$$\frac{d^2\theta_o}{dt^2} = 8\theta_i - 1.6\frac{d\theta_o}{dt} - 8\theta_o$$

From this equation we determine the gains needed to simulate the system. The highest-order derivative tells us the number of integrations needed to complete the simulation of the system. In this case two are needed. From the constant coefficients of the differential equation, the gains needed are 8, -1.6, and -8. The diagram of this system is shown in Figure 8-7. The input for this analysis is an impulse.

The damping ratio for this circuit is 0.282. The undamped natural frequency, ω_n, of the system is 2.828 rad/s. The netlist for this analysis is

```
*MASS INERTIA DEMO CH. 8
VPUL 1 0 PULSE(0 1 0 10U 10U 1 8)
R1 1 2 125K
R2 7 2 125K
R3 2 3 625K
R4 3 4 1MEG
R5 5 6 1MEG
```

```
R6 6 7 1MEG
C1 2 3 1U
C2 4 5 1U
XAMP1 0 2 3 LF155
XAMP2 0 4 5 LF155
XAMP3 0 6 7 LF155
.TRAN .1 5 0 .01 UIC
.IC V(3)=0 V(5)=0
.PROBE
.LIB SUB.LIB
.END
```

The output of the analysis is shown in Figure 8-8. The damping ratio is controlled by the feedback of $-dx/dt$, the damping ratio can be any desired value by changing the gain for this input. You should change the gain and rerun the analysis.

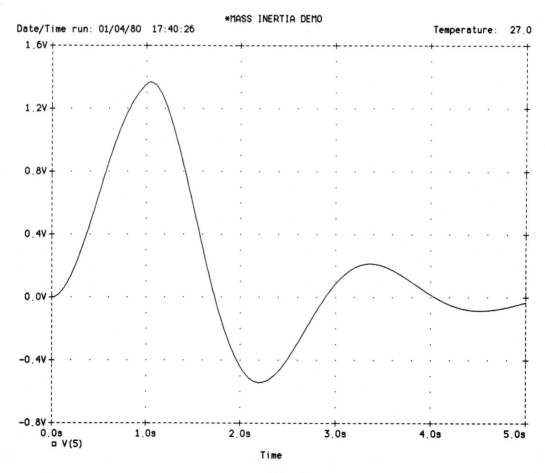

Figure 8-8

8.3. UNIVERSAL DAMPING CURVES

The underdamped condition in a feedback control system implies some instability of the system. The amount of instability is a function of the damping ratio of the system. Books that treat the subject of feedback control systems publish graphs of the damping ratio of second-order systems. These graphs are usually normalized and are then usable for any type of system. Usually, these graphs are small and do not include the specific damping ratio needed. With PSpice, it is a simple matter to construct a damping chart for any damping ratio. You can also create a group of damping charts for any range of damping ratios. You can then normalize them and then have larger, easier-to-read charts. A similar analysis was done in Chapter 5. We can simply make the same type of netlist, and then clone the list a few times to get all the charts we may need. The equation for the required resistance is

$$R = 2\zeta \sqrt{\frac{L}{C}}$$

Let us make a chart for ζ values of 0.2 to 1.0. The circuit for the analysis is shown in Figure 8-9.

```
*DAMPING CHART FOR ζ = 0.2
V1 1 0 PWL(0,0 1U,1 14.999M,1V 15M,0)
R1 1 2 40
L1 2 3 .1
C1 3 0 1E-5
.TRAN .025M 15M 0 0.05M
.PROBE
.END
*DAMPING CHART FOR ζ = 0.4
V1 1 0 PWL(0,0 1U,1 14.999M,1V 15M,0)
R1 1 2 80
L1 2 3 .1
C1 3 0 1E-5
.TRAN .025M 15M 0 0.05M
.PROBE
.END
```

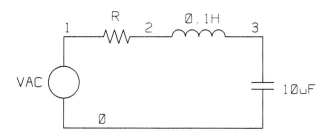

Figure 8-9

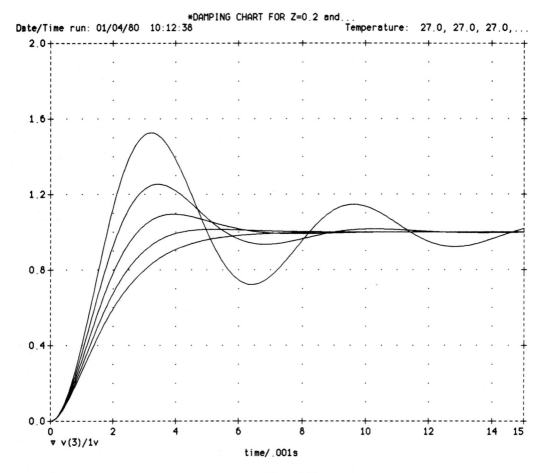

Figure 8-10

```
*DAMPING CHART FOR ζ = 0.6
V1  1  0  PWL(0,0 1U,1 14.999M,1V 15M,0)
R1  1  2  120
L1  2  3  .1
C1  3  0  1E-5
.TRAN .025M 15M 0 0.05M
.PROBE
.END
*DAMPING CHART FOR ζ = 0.8
V1  1  0  PWL(0,0 1U,1 14.999M,1V 15M,0)
R1  1  2  160
L1  2  3  .1
C1  3  0  1E-5
```

```
.TRAN .025M 15M 0 0.05M
.PROBE
.END
*DAMPING CHART FOR ζ = 1.0
V1 1 0 PWL(0,0 1U,1 14.999M,1V 15M,0)
R1 1 2 200
L1 2 3 .1
C1 3 0 1E-5
.TRAN .025M 15M 0 0.05M
.PROBE
.END
```

The composite output of this analysis is shown in Figure 8-10. The charts shown are normalized and can be used for any type of second-order system.

SUMMARY

PSpice can be used to analyze both simple and complex feedback control systems in both the frequency and time domains. PSpice can be used to generate plots that can be used with Nichols charts. PSpice can be used to generate a chart for any damping ratio.

SELF-EVALUATION

1. Analysis of plant response is done using the circuit of Figure P8-1. Run an analysis of this circuit. The analysis should run for about 2s. The plant transfer function is $3/(s^2 + 8s + 12)$.

2. Derivative control is used to decrease the response time of a controller. This is shown in Figure P8-2. Analyze the system to show that the response time is faster than the plant alone.

3. To make a system have zero error, the integral controller is used. This is shown in Figure P8-3. The response time of this controller is much longer than the derivative controller. Analyze the controller to show the zero error and longer response time.

4. The best available controllers use proportional, integral, and derivative control in what is called a PID controller. This type of controller is shown in Figure P8-4. Analyze this controller to show that the response settles within 0.5 s and that the error is zero.

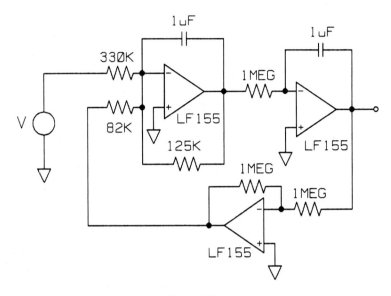

Figure P8-1

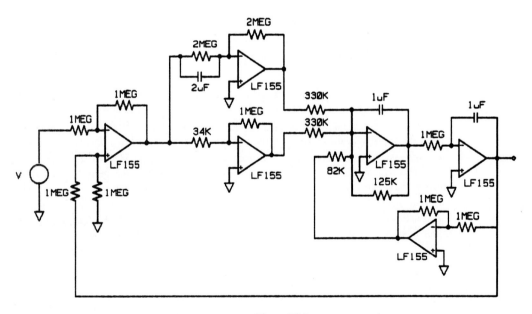

Figure P8-2

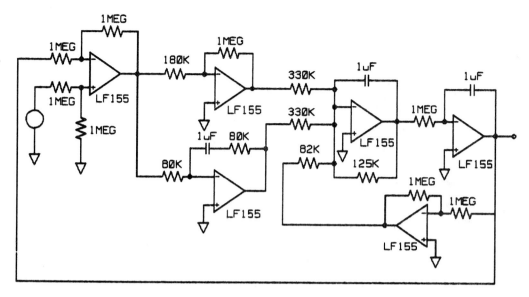

Figure P8-3

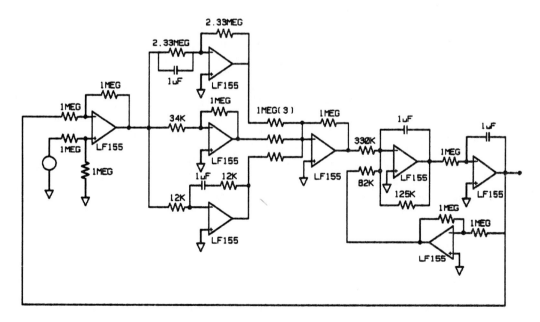

Figure P8-4

$$\boxed{ \Large 9 }$$

Noise Calculations Using PSpice

OBJECTIVES

1. To learn to use the noise calculation abilities of PSpice.
2. To apply the technique of noise analysis to both active and passive circuits.

INTRODUCTION

All electronic devices generate noise. Noise can be defined as any undesired signal. To be specific, the noise we are concerned with in this chapter is thermal noise and shot noise. This noise is generated by devices in circuits. Of the two types of noise, shot noise has the larger amplitude and therefore is the more troublesome. Thermal noise is the noise generated in resistive elements. Unless the resistor is large in value, the amount of thermal noise is usually small. The noise generated is in the form of a voltage or current. This being true, it is treated as a real signal by the circuit, giving an output with no input signal.

9.1. THERMAL NOISE

Thermal noise is found in resistive elements and is usually given either as V^2/Hz or $V/\sqrt{\text{Hz}}$. The amount of noise voltage and current generated in a resistance is

$$V_t = \sqrt{4kTR\,\Delta f}$$
$$I_t = \sqrt{4kT\,\Delta f/R}$$

where k is Boltzmann's constant, 1.38×10^{-23}; T is the absolute temperature in Kelvin; R is the resistance in ohms; and Δf is the bandwidth of the circuit in hertz. The voltage formula produces the voltage across the resistance directly, but requires the addition of a node to the circuit. For this reason, the current equation is used since it is across the resistor.

9.2. SHOT NOISE

Shot noise is associated with semiconductor junctions. It is a function of the current in the device and is given by

$$I_s^2 = 2qI_{dc}\, \Delta f$$

where q is the electronic charge, 1.602×10^{-19} C, and I_{dc} is the dc current in the device.

9.3. NOISE CALCULATIONS IN PSPICE

Noise analysis in PSpice requires a frequency response analysis to be done also. The noise is calculated for each of the frequencies in the analysis. The syntax for the noise analysis is

```
.NOISE V( <node> [,<node>]) <source name> [interval]
```

The voltage nodes are where the total output noise is taken. The nodes may be a single node referencing against ground or a pair of nodes across which the output noise is measured. The source is an independent voltage or current source. The total output noise is referred to the source to calculate the equivalent input noise. When [interval] is specified, the noise calculations are printed to a table. The table is generated without the need for .PRINT or .PLOT statements.
There are four parameters available from a noise analysis:

1. ONOISE   total noise at the specified output
2. INOISE   ONOISE referred to the input source
3. DB(ONOISE)   ONOISE in dB (compared to 1 v/$\sqrt{\text{Hz}}$)
4. DB(INOISE)   INOISE in dB (compared to 1 v/$\sqrt{\text{Hz}}$)

These quantities may be printed out using the .PRINT or .PLOT command. The syntax for the .PRINT or .PLOT command is

```
.PRINT NOISE <output 1> <output 2> · · ·
.PLOT NOISE <output 1> <output 2> · · ·
```

Each output value will be a column in the output table or a curve on a graph. If [interval] is used, a detailed table of noise outputs will be generated independently of the .PRINT table. The frequencies used in the table will be those set by the interval used.

9.4. *RC* INTEGRATOR NOISE CALCULATIONS

To demonstrate the use of the noise analysis, a *RC* low-pass filter will be analyzed. This circuit is not usually used in an area where noise would be a concern. It is used here to correlate the values of the noise voltage output and the filter transfer function. You should know that thermal noise is *white noise*. This means that the noise spectrum

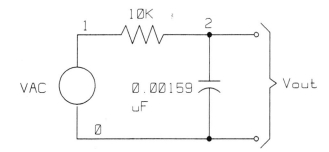

Figure 9-1

has a constant density. We shall use the [interval] part of the command to print the output noise information for every tenth frequency. You will see that the transfer function is calculated for the filter at each of the frequencies. The capacitor contributes no noise, and the noise spectrum of the resistor is constant with frequency. For these reasons the noise output of the filter depends on the transfer function only. The circuit for the filter is shown in Figure 9-1, and the netlist for the analysis is

```
*RC INTEGRATOR NOISE DEMO
VAC 1 0 AC 1
R1 1 2 100K
C1 2 0 .00159U
.AC DEC 10 1 1MEG
.NOISE V(2) VAC 10
.PROBE
.END
```

The output analysis using PROBE is shown in Figure 9-2. The dc output file is shown in Analysis 9-1. You can see immediately that the noise output of the filter is the noise voltage of the resistor multiplied by the transfer function. There are some other important parameters that we can calculate for the filter to give us more information. We can use PROBE's integration capability to sum the values of all the noise voltages to find the total noise of the system. This is done by summing the squares of

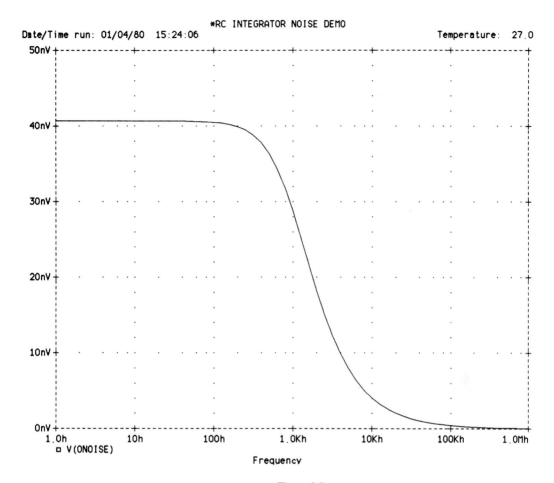

Figure 9-2

all the noise voltages calculated, then taking the square root of the sum. The syntax for the calculation is

$$SQRT(S(V(ONOISE)*V(ONOISE)))$$

The output of this calculation is shown in Figure 9-3.

The signal-to-noise ratio of the system can also be shown using PROBE. This will be shown in the next analysis.

```
******* 01/04/80 ******* Evaluation PSpice (January 1990) ******* 15:24:06 *******
*RC INTEGRATOR NOISE DEMO
****      NOISE ANALYSIS                    TEMPERATURE =    27.000 DEG C
***********************************************************************
     FREQUENCY =  1.000E+03 HZ
**** RESISTOR SQUARED NOISE VOLTAGES (SQ V/HZ)
               R1
TOTAL     8.296E-16
**** TOTAL OUTPUT NOISE VOLTAGE          =  8.296E-16 SQ V/HZ
                                         =  2.880E-08 V/RT HZ

     TRANSFER FUNCTION VALUE:
        V(2)/VAC                 =  7.075E-01
     EQUIVALENT INPUT NOISE AT VAC =  4.071E-08 V/RT HZ
******* 01/04/80 ******* Evaluation PSpice (January 1990) ******* 15:24:06 *******
*RC INTEGRATOR NOISE DEMO
****      NOISE ANALYSIS                    TEMPERATURE =    27.000 DEG C
***********************************************************************
     FREQUENCY =  1.000E+04 HZ
**** RESISTOR SQUARED NOISE VOLTAGES (SQ V/HZ)
               R1
TOTAL     1.644E-17
**** TOTAL OUTPUT NOISE VOLTAGE          =  1.644E-17 SQ V/HZ
                                         =  4.055E-09 V/RT HZ

     TRANSFER FUNCTION VALUE:
        V(2)/VAC                 =  9.960E-02
     EQUIVALENT INPUT NOISE AT VAC =  4.071E-08 V/RT HZ
```

Analysis 9-1

```
******* 01/04/80 ******* Evaluation PSpice (January 1990) ******* 15:24:06 *******
*RC INTEGRATOR NOISE DEMO
****     NOISE ANALYSIS                    TEMPERATURE =   27.000 DEG C
**********************************************************************************
     FREQUENCY =  1.000E+01 HZ
**** RESISTOR SQUARED NOISE VOLTAGES (SQ V/HZ)
              R1
TOTAL    1.657E-15
**** TOTAL OUTPUT NOISE VOLTAGE        =  1.657E-15 SQ V/HZ
                                       =  4.071E-08 V/RT HZ

     TRANSFER FUNCTION VALUE:
       V(2)/VAC                  =  1.000E+00
     EQUIVALENT INPUT NOISE AT VAC =  4.071E-08 V/RT HZ
******* 01/04/80 ******* Evaluation PSpice (January 1990) ******* 15:24:06 *******
*RC INTEGRATOR NOISE DEMO
****     NOISE ANALYSIS                    TEMPERATURE =   27.000 DEG C
**********************************************************************************
     FREQUENCY =  1.000E+02 HZ
**** RESISTOR SQUARED NOISE VOLTAGES (SQ V/HZ)
              R1
TOTAL    1.641E-15
**** TOTAL OUTPUT NOISE VOLTAGE        =  1.641E-15 SQ V/HZ
                                       =  4.051E-08 V/RT HZ

     TRANSFER FUNCTION VALUE:
       V(2)/VAC                  =  9.950E-01
     EQUIVALENT INPUT NOISE AT VAC =  4.071E-08 V/RT HZ
```

Analysis 9-1 (cont'd)

```
******* 01/04/80 ******* Evaluation PSpice (January 1990) ******* 15:24:06 *******
*RC INTEGRATOR NOISE DEMO

****      NOISE ANALYSIS                    TEMPERATURE =    27.000 DEG C

**********************************************************************************

     FREQUENCY =  1.000E+05 HZ

**** RESISTOR SQUARED NOISE VOLTAGES (SQ V/HZ)

             R1

TOTAL      1.661E-19

**** TOTAL OUTPUT NOISE VOLTAGE          = 1.661E-19 SQ V/HZ

                                         = 4.075E-10 V/RT HZ

     TRANSFER FUNCTION VALUE:

        V(2)/VAC                 = 1.001E-02

     EQUIVALENT INPUT NOISE AT VAC = 4.071E-08 V/RT HZ

******* 01/04/80 ******* Evaluation PSpice (January 1990) ******* 15:24:06 *******
*RC INTEGRATOR NOISE DEMO

****      NOISE ANALYSIS                    TEMPERATURE =    27.000 DEG C

**********************************************************************************

     FREQUENCY =  1.000E+06 HZ

**** RESISTOR SQUARED NOISE VOLTAGES (SQ V/HZ)

             R1

TOTAL      1.661E-21

**** TOTAL OUTPUT NOISE VOLTAGE          = 1.661E-21 SQ V/HZ

                                         = 4.075E-11 V/RT HZ

     TRANSFER FUNCTION VALUE:

        V(2)/VAC                 = 1.001E-03

     EQUIVALENT INPUT NOISE AT VAC = 4.071E-08 V/RT HZ

        JOB CONCLUDED

        TOTAL JOB TIME          5.99
```

Analysis 9-1 (cont'd)

```
******* 01/04/80 ******* Evaluation PSpice (January 1990) ******* 15:24:06 *******

*RC INTEGRATOR NOISE DEMO CHAPTER 9 ANALYSIS 9-1

****       CIRCUIT DESCRIPTION

**********************************************************************************

VAC 1 0 AC 1
R1 1 2 100K
C1 2 0 .00159U
.AC DEC 10 1 1MEG
.NOISE V(2) VAC 10
.PROBE
.END

******* 01/04/80 ******* Evaluation PSpice (January 1990) ******* 15:24:06 *******

*RC INTEGRATOR NOISE DEMO

****       SMALL SIGNAL BIAS SOLUTION        TEMPERATURE =    27.000 DEG C

**********************************************************************************

 NODE    VOLTAGE      NODE    VOLTAGE      NODE    VOLTAGE      NODE    VOLTAGE

(   1)    0.0000  (    2)    0.0000

    VOLTAGE SOURCE CURRENTS
    NAME            CURRENT

    VAC          0.000E+00

    TOTAL POWER DISSIPATION    0.00E+00   WATTS

******* 01/04/80 ******* Evaluation PSpice (January 1990) ******* 15:24:06 *******

*RC INTEGRATOR NOISE DEMO

****       NOISE ANALYSIS                 TEMPERATURE =    27.000 DEG C

**********************************************************************************

    FREQUENCY =  1.000E+00 HZ

**** RESISTOR SQUARED NOISE VOLTAGES (SQ V/HZ)

           R1

TOTAL    1.658E-15

**** TOTAL OUTPUT NOISE VOLTAGE          =  1.658E-15 SQ V/HZ

                                         =  4.071E-08 V/RT HZ
    TRANSFER FUNCTION VALUE:

    V(2)/VAC                  =  1.000E+00

    EQUIVALENT INPUT NOISE AT VAC =  4.071E-08 V/RT HZ
```

Analysis 9-1 (cont'd)

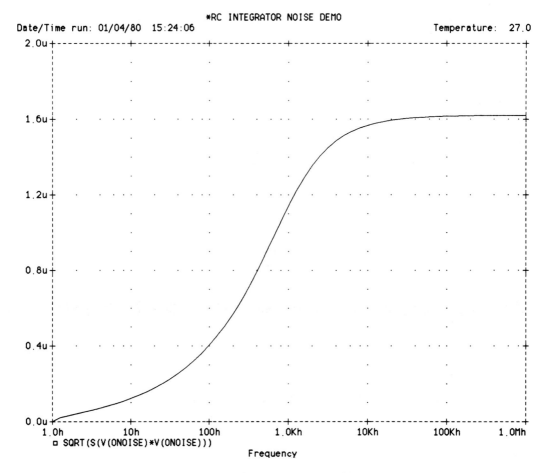

Figure 9-3

9.5. NOISE CALCULATIONS FOR A SINGLE-STAGE AMPLIFIER

The next example shows the noise calculations for a simple single-stage amplifier. Once again, there must be an ac sweep done, and the noise will be calculated for each of the frequencies of the sweep. The circuit for the amplifier is shown in Figure 9-4 and the netlist for the amplifier is

```
*2N3904 AMPLIFIER NOISE
VCC 1 0 15
VAC 2 0 AC 1
R1 1 4 1.2MEG
R2 1 3 4K
CNOISE 3 0 50P
```

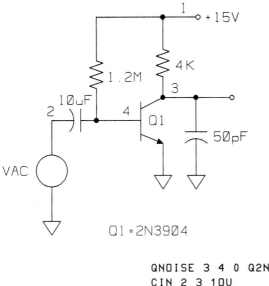

Figure 9-4

```
QNOISE 3 4 0 Q2N3904
CIN 2 3 10U
.AC DEC 25 1 10MEG
.LIB
.PROBE
.NOISE V(3) VAC 25
.END
```

A portion of the output file for the noise calculations is printed below. This portion of the printout has been condensed to show the noise outputs for 1 Hz, 10 kHz, and 1 MHz.

```
      FREQUENCY = 1.000E + 00 HZ
**** TRANSISTOR SQUARED NOISE VOLTAGES (SQ V/HZ)
          QNOISE
 RB       2.595E-16
 RC       1.158E-22
 RE       0.000E+00
 IB       1.516E-12
 IC       7.947E-15
 FN       0.000E+00
 TOTAL    1.524E-12
**** RESISTOR SQUARED NOISE VOLTAGES (SQ V/HZ)
            R1            R2
 TOTAL    5.477E-15   5.568E-17
**** TOTAL OUTPUT NOISE VOLTAGE        = 1.530E-12 SQ V/HZ
                                       = 1.237E-06 V/RT HZ
          TRANSFER FUNCTION VALUE:
            V(3)/VAC                   = 3.956E+01
          EQUIVALENT INPUT NOISE AT VAC = 3.126E-08 V/RT HZ
```

```
        FREQUENCY = 1.000E+04 HZ
**** TRANSISTOR SQUARED NOISE VOLTAGES (SQ V/HZ)
           QNOISE
   RB       1.071E-14
   RC       1.158E-22
   RE       0.000E+00
   IB       2.533E-17
   IC       7.945E-15
   FN       0.000E+00
   TOTAL    1.868E-14
**** RESISTOR SQUARED NOISE VOLTAGES (SQ V/HZ)
                 R1          R2
   TOTAL     2.260E-21   5.567E-17
**** TOTAL OUTPUT NOISE VOLTAGE       = 1.873E-14 SQ V/HZ
                                      = 1.369E-07 V/RT HZ


        TRANSFER FUNCTION VALUE:
           V(3)/VAC                   = 2.542E+02
        EQUIVALENT INPUT NOISE AT VAC = 5.385E-10 V/RT HZ


        FREQUENCY = 1.000E+06 HZ
**** TRANSISTOR SQUARED NOISE VOLTAGES (SQ V/HZ)
           QNOISE
   RB       4.301E-15
   RC       7.837E-23
   RE       0.000E+00
   IB       9.922E-18
   IC       3.191E-15
   FN       0.000E+00
   TOTAL    7.502E-15
**** RESISTOR SQUARED NOISE VOLTAGES (SQ V/HZ)
                 R1          R2
   TOTAL     9.078E-26   2.236E-17
**** TOTAL OUTPUT NOISE VOLTAGE       = 7.524E-15 SQ V/HZ
                                      = 8.674E-08 V/RT HZ
        TRANSFER FUNCTION VALUE:
           V(3)/VAC                   = 1.611E+02
        EQUIVALENT INPUT NOISE AT VAC = 5.385E-10 V/RT HZ
```

In this analysis, flicker noise has been ignored. This type of noise is usually a very-low-frequency phenomenon. It is a problem only near dc or at most a few hertz. This type of noise is often called 1/f noise.

All the noise voltages shown are referenced to the output of the amplifier. If you need to know the individual noise voltages, divide the desired noise voltage by the square of the gain of the amplifier. The output noise total is the sum of the transistor and resistor noise voltages. This is given as volts squared per hertz. The total input noise voltage is simply this value divided by the square of the gain given by the transfer function.

It is possible to obtain the total output noise over the bandwidth of the amplifier by using PROBE and the cursor function. This is done using the command

$$\texttt{SQRT(S(V(ONOISE)*V(ONOISE)))}$$

The noise shown on this graph is a running total of the noise at any frequency. It is the root-mean-square noise voltage over the frequencies measured. The bandwidth of this amplifier is about 800 kHz, and the total noise is about 100 μV over the bandwidth. The plot of the noise voltage versus frequency is shown in Figure 9-5.

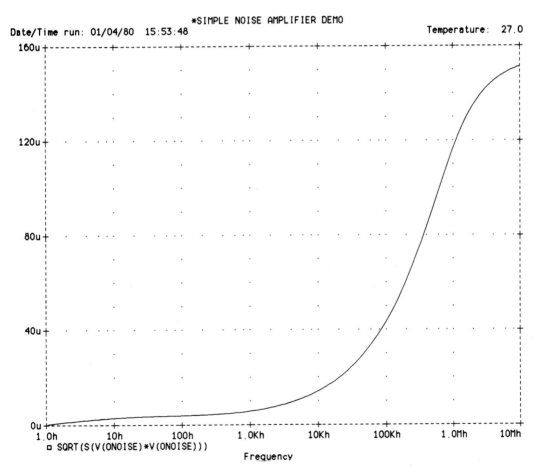

Figure 9-5

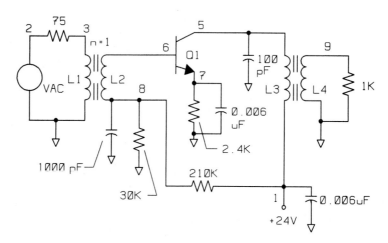

Figure 9-6

9.6. CALCULATING THE SIGNAL-TO-NOISE RATIO OF AN RF AMPLIFIER

For this example a single-stage RF amplifier will be analyzed. The signal-to-noise ratio is important in this amplifier because the input to this amplifier can be nearly the same level as the noise. This puts a lower limit on the signal input that can be amplified. If this amplifier were the first stage in a radio, it would be the primary determiner of the noise figure of the system. As before, the output noise of the amplifier is limited to the bandwidth of the amplifier. The circuit for this amplifier is shown in Figure 9-6. This amplifier is operating at a center frequency of 5 MHz and the RF generator input is 10 μV. The Q value of the output tuned circuit is about 50. The netlist for the amplifier is

```
*RF AMPLIFIER NOISE
VCC 1 0 24
VAC 2 0 AC 10U
R1 3 0 75
L1 3 0 1.19U
L2 6 8 1.19U
K12 L1 L2 .9999
R2 8 0 30K
R2 8 1 210K
R3 7 0 2.4K
C1 8 0 1000P
C2 7 0 .006U
C3 1 0 .006U
CT 5 0 100P
L3 5 4 10U
```

```
L4 9 0 1U
K34 L3 L4 .9999
Q1 5 6 7 Q2N3904
RL 9 0 1000
.LIB
.AC DEC 300 1MEG 10MEG
.NOISE V(9) VAC 20
.PROBE
.END
```

A plot of the output of this amplifier is shown in Figure 9-7. In this figure the output produced by both the noise and the signal are displayed. Note that with a 10-μV input signal, the output gives a noise figure of about 24 dB for this amplifier. The signal-to-noise ratio at the output of an amplifier is

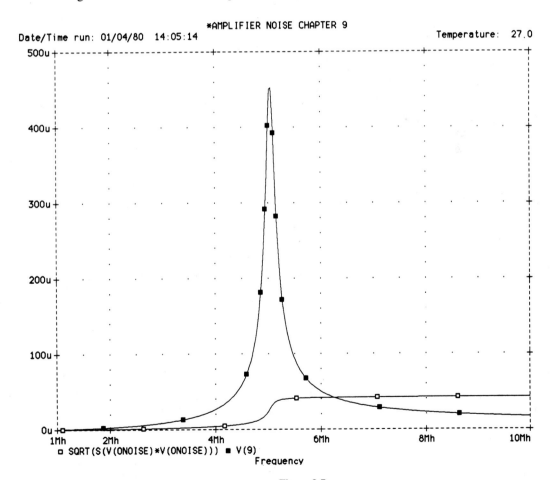

Figure 9-7

$$dB = 20 \times \log \frac{V_{out(signal)}}{V_{out(noise)}}$$

which for PROBE is

$$DB(Vout/SQRT(S(V(ONOISE)*V(ONOISE))))$$

The plot of this calculation is shown in Figure 9-8.

Any of the noise calculations for a circuit can be done using PROBE. All that is required is to transform the calculations into a form suitable for PSpice. You could, for instance, calculate the noise figure for the amplifier using the formula

$$NF = 20 \log \frac{V_{in(signal)}/V_{in(noise)}}{V_{out(signal)}/V_{out(noise)}}$$

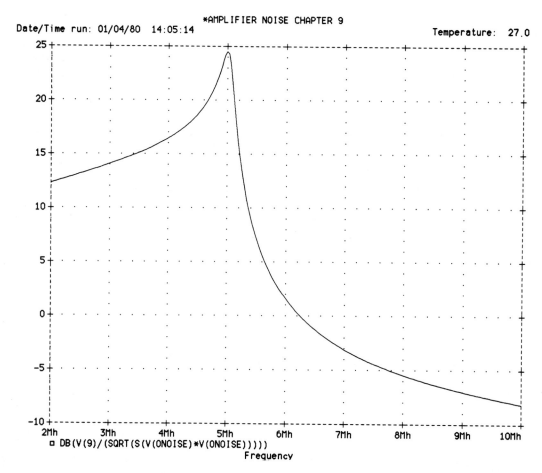

Figure 9-8

Although this is somewhat cumbersome, it will fit on one line which is the determining criteria for most formulas used with PROBE.

SUMMARY

PSpice can do both thermal and shot noise calculations for devices. To do a noise analysis, a frequency response analysis is required.

PSpice will automatically print the results of a noise analysis to the dc output file if the interval for the noise measurement is specified. PSpice can also be used to calculate the total noise in an active or passive circuit.

SELF-EVALUATION

1. For the simple LCR circuit shown in Figure P9-1, produce the output noise voltage curves.
2. For the circuit shown in Figure P9-2 determine the input and output rms noise voltage with frequency. Determine the signal-to-noise ratio of the circuit.
3. For the circuit shown in Figure P9-3, determine the total output noise over the audio range of 20 to 20000 Hz.
4. For the circuit shown in Figure P9-4, determine the input and output rms noise voltage and the noise factor of the amplifier. (*Note:* the *y* axis will need to be rescaled to show this.) Use the following function to determine the noise figure:

```
DB(V(2)/SQRT(S(V(INOISE)*V(INOISE)))-V(7)/
SQRT(S(V(ONOISE*V(ONOISE)))))
```

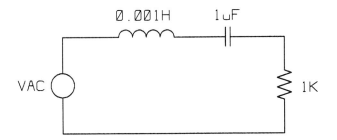

Figure P9-1

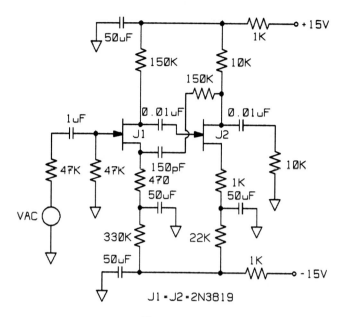

Figure P9-2

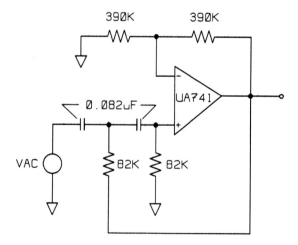

Figure P9-3

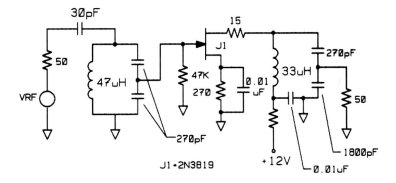

Figure P9-4

10

Power Amplifiers

OBJECTIVE

1. To learn to use PSpice to analyze several types of power amplifier.

INTRODUCTION

Power amplifiers are used in all areas of electronics. From audio to servomechanisms, power amplifiers are the workhorses in any circuit. In this chapter we investigate some of the methods of analyzing power amplifiers using PSpice.

10.1. CLASS A POWER AMPLIFIER ANALYSIS

The class A power amplifier is the least efficient of the types of power amplifiers available. For this analysis we analyze a transformer coupled class A amplifier that delivers 50 mW to an 8-Ω load resistor. This is similar to what you may find in a cheap transistor radio. Both ac, transient, and Fourier analysis will be done so that we may find both frequency response and distortion of the amplifier. The transient analysis also allows us to check the clipping levels of the amplifier, and the voltages anywhere in the circuit. If the amplifier does not clip symmetrically, the bias for the amplifier is probably wrong. The transistor used in this amplifier is a 2N3904. While the output power

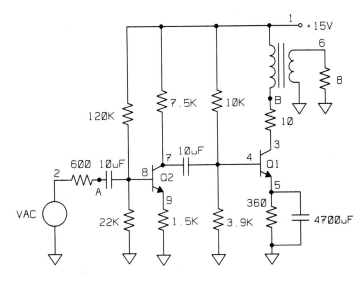

Figure 10-1

of this amplifier is small, the analysis is applicable to any power level. The circuit of the amplifier is shown in Figure 10-1. The netlist for the amplifier is

```
*CHAPTER 10 PROGRAM 1 - CLASS A AMPLIFIER
VCC 1 0 15
VAC 2 0 AC 1 SIN(0 .1 1K)
R1 1 4 10K
R2 4 0 3.9K
R3 5 0 360
RL 6 0 8
RG 2 A 600
RD 3 B 10
C1 7 4 10U
C2 5 0 4700U
L1 1 B 1
L2 6 0 .00444
K12 L1 L2 .9999
R4 1 7 7.5K
R5 9 0 1.5K
R6 1 8 120K
R7 8 0 22K
CIN A 8 10U
Q1 3 4 5 Q2N3904
Q2 7 8 9 Q2N3904
.AC DEC 20 1 1MEG
.TRAN .01M .005 0 .025M
```

```
.PROBE
.FOUR 1000 V(3) V(6)
.OP
.LIB
.END
```

The frequency response of this amplifier is shown in Figure 10-2, and the transient response is shown in Figure 10-3. Note the narrow frequency response of the amplifier. The low-end frequency response suffers most because the transformer inductance is small. The theoretical maximum efficiency of class A amplifiers is 50%. Analysis of the output of this amplifier shows that it is operating at about 30%. This is what most well-designed class A amplifiers will perform at. The Fourier analysis is in the output

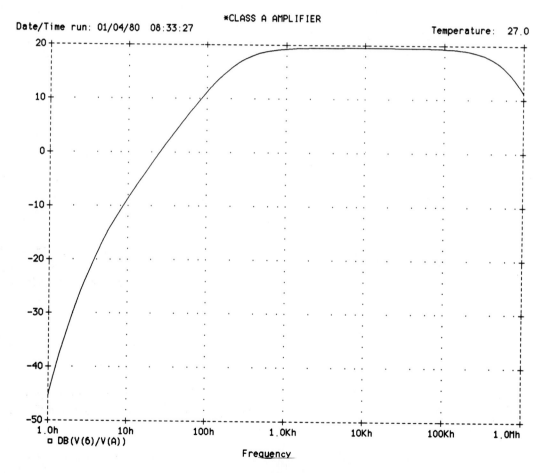

Figure 10-2

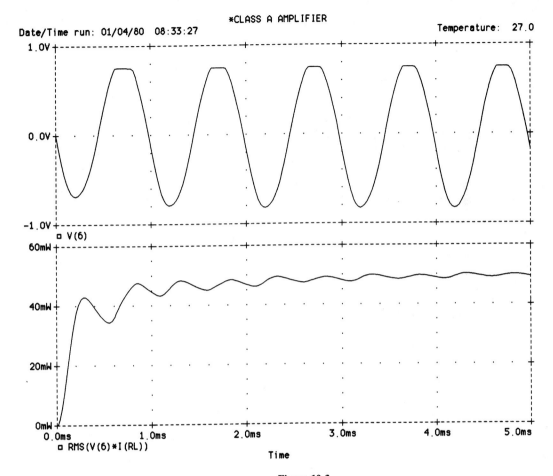

Figure 10-3

file of the amplifier. Even though the output waveform shows distortion, the analysis shows that it is only about 5% for this amplifier.

10.2. CLASS B POWER AMPLIFIERS

The most popular of the types of amplifier for use in audio and other types of systems is the class B transformerless amplifier. This type of amplifier is usually of two different types:

1. Complementary symmetry type
2. Quasi-complementary symmetry type

There are only two important differences between the two types of amplifier. The complementary symmetry amplifier often uses two power supplies. It also uses a complementary PNP–NPN transistor pair in the output. The output devices are usually Darlington transistors. The output devices are usually connected as emitter followers. The quasi-complementary amplifier uses only a single supply and the same type of transistor, usually NPN, for the output. In the quasi-complementary amplifier, one of the output transistors is an emitter follower; the other is a collector follower.

Both types of amplifier are similar in power output capability. Both suffer an effect called crossover distortion. Crossover is the result of the base–emitter diode voltage drop. This type of distortion is eliminated by applying a small amount of dc bias when no signal is applied. The bias is applied by the use of a current mirror.

We will analyze one of each of the two types of amplifier. The first is a 30-W amplifier of the complementary symmetry type. For this analysis we will make two Darlington transistors and use them as subcircuits from a library. The netlist for these two Darlington transistors is

```
.SUBCKT QDAR1 1 4 3
Q1 1 2 3 QMOD1 20
Q2 4 1 2 QMOD2
.MODEL QMOD1 NPN(BF=100 CJC=3PF CJE=5PF VA=150)
.MODEL QMOD2 NPN(BF=150 CJC=1PF CJE=2PF VA=100)
.ENDS


.SUBCKT QDAR2 10 40 30
Q1 10 20 30 QMOD3 20
Q2 40 10 20 QMOD4
.MODEL QMOD3 PNP(BF=100 CJC=3PF CJE=5PF VA=150)
.MODEL QMOD4 PNP(BF=150 CJC=1PF CJE=2PF VA=100)
.ENDS.
```

The theoretical voltage supply required for this type of amplifier is calculated from

$$V_{cc} = \sqrt{8P_oR_L}$$

This formula does not take into account transistor parameters that may cause less than ideal saturation or other effects to occur. For this reason it is wise to make the power supply this value and add at least 10% or more to the voltage. You can always back the power supply requirements down. The required voltage for this amplifier is then

$$V_{cc} = \sqrt{8(8)(30)} = 43.81 \text{ V}$$

For the sake of a working amplifier, let us make this 50 V. This will handle any saturation problems or problems of similar type that come up.

The circuit for the amplifier is shown in Figure 10-4. The netlist for the amplifier is

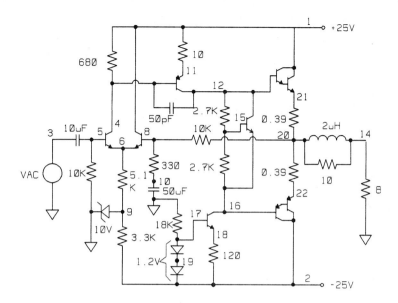

Figure 10-4

```
*COMPLEMENTARY SYMMETRY AMPLIFIER
VCC 1 0 25
VEE 2 0 -25
VAC 3 0 AC 1 SIN(0 .7 1K)
R1 5 0 10K
R2 4 1 680
R3 6 9 5.1K
R4 9 2 3.3K
R5 8 10 330
R6 8 20 10K
R7 17 0 18K
R8 12 15 2.7K
R11 15 16 2.7K
R12 18 2 120
R13 21 20 .39
R14 22 20 .39
R15 20 14 10
R16 1 11 10
RL 14 0 8
C1 3 5 10U
C2 10 0 50U
C3 12 4 50P
L1 20 14 2U
Q1 4 5 6 Q2N3904
Q2 1 8 6 Q2N3904
Q3 12 4 11 Q2N3906
Q4 12 15 16 Q2N3904
```

```
Q5 16 17 18 Q2N3904
XDAR1 1 12 21 QDAR1
XDAR2 2 16 22 QDAR2
D1 9 0 ZMOD
D2 17 19 DMOD1
D3 19 2 DMOD1
.MODEL DMOD1 D
.MODEL ZMOD D(BV=10)
.LIB NOM.LIB
.LIB SUB.LIB
.TRAN .05M .0025 0 .025M
.FOUR 1K V(20)
.AC DEC 25 1 1MEG
.PROBE
.END
```

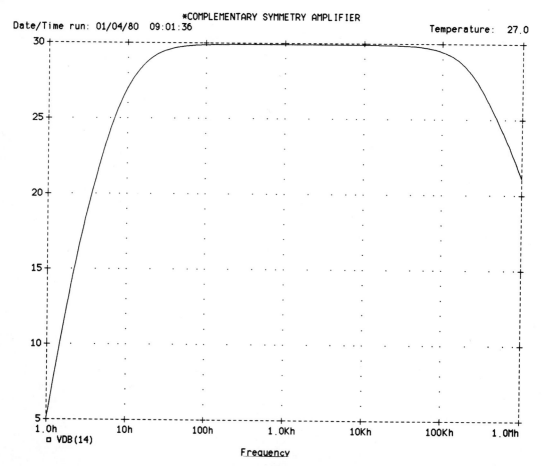

*COMPLEMENTARY SYMMETRY AMPLIFIER

Date/Time run: 01/04/80   09:01:36   Temperature: 27.0

□ VDB(14)

Frequency

Figure 10-5

The frequency response of this amplifier is from about 10 Hz to 125 kHz, and the output power is about 35 W with 0.7 V peak input. The output frequency response is shown in Figure 10-5, and the rms power output is shown in Figure 10-6. The output file of the amplifier shows the distortion at 1 kHz to be about 0.3%. The Fourier analysis portion of the output file is shown in Analysis 10-1.

The last of the class B amplifiers we shall analyze is the quasi-complementary symmetry amplifier. The amplifier is a simple 0.5-W output amplifier that uses several configurations of amplifiers. The input stage is a common-emitter amplifier of moderate gain. The outputs are a Darlington pair and a Sziklai pair. The Sziklai pair acts as a very-high-beta NPN transistor. This type of amplifier circuit is fairly common in the industry and is also found in integrated circuit power amplifiers. The circuit of this amplifier is shown in Figure 10-7.

The netlist for the amplifier is

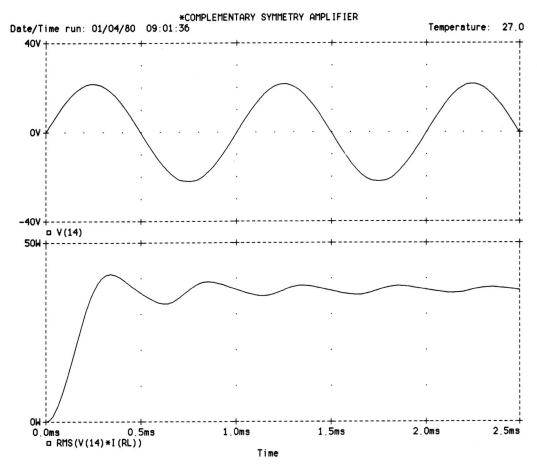

Figure 10-6

```
******* 01/04/80 ******* Evaluation PSpice (January 1990) ******* 09:01:36 *******

*COMPLEMENTARY SYMMETRY AMPLIFIER

****      FOURIER ANALYSIS                 TEMPERATURE =    27.000 DEG C

'***********************************************************************

FOURIER COMPONENTS OF TRANSIENT RESPONSE V(20)

DC COMPONENT =  -2.473243E-01

HARMONIC   FREQUENCY    FOURIER     NORMALIZED    PHASE       NORMALIZED
  NO         (HZ)      COMPONENT    COMPONENT     (DEG)      PHASE (DEG)
   1       1.000E+03   2.192E+01    1.000E+00    -1.794E+02    0.000E+00
   2       2.000E+03   1.535E-02    7.003E-04    -9.204E+01    8.738E+01
   3       3.000E+03   3.414E-02    1.557E-03    -1.768E+02    2.644E+00
   4       4.000E+03   3.087E-02    1.408E-03     9.400E+01    2.734E+02
   5       5.000E+03   2.703E-02    1.233E-03     3.712E+00    1.831E+02
   6       6.000E+03   2.950E-02    1.346E-03    -9.012E+01    8.930E+01
   7       7.000E+03   2.694E-02    1.229E-03    -1.753E+02    4.153E+00
   8       8.000E+03   2.526E-02    1.152E-03     9.090E+01    2.703E+02
   9       9.000E+03   2.005E-02    9.147E-04    -6.428E-01    1.788E+02


     TOTAL HARMONIC DISTORTION =    3.449721E-01 PERCENT


          JOB CONCLUDED

          TOTAL JOB TIME       418.97
```

Analysis 10-1

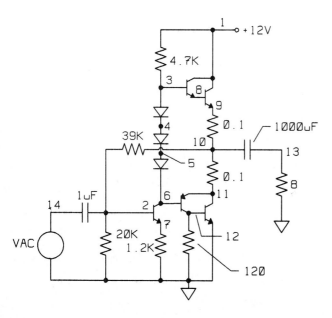

Figure 10-7

```
*QUASI-COMPLEMENTARY AMPLIFIER-CHAPTER 10
VCC 1 0 12
VAC 14 0 AC 1 SIN(0 0.8 1K)
R1  2 1 39K
R2  2 0 20K
R3  1 3 4.7K
R4  7 0 1.2K
R7  10 9 .1
R8  10 11 .1
RL  13 0 8
C1  14 2 1U
C2  10 13 1000U
D1  3 4 D1N4148
```

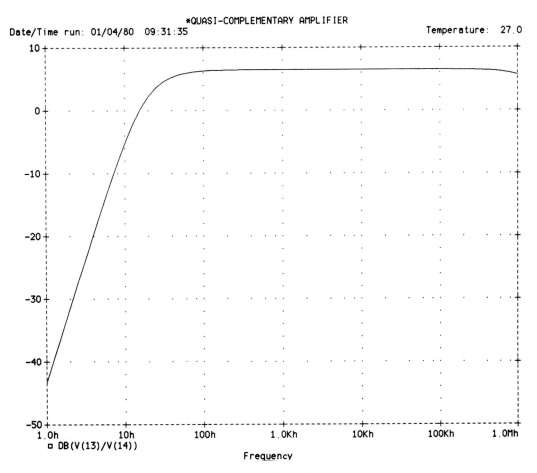

Figure 10-8

```
D2 4 5 D1N4148
D3 5 6 D1N4148
Q1 6 2 7 Q2N3904
Q2 1 3 8 Q2N3904
Q3 12 6 11 Q2N3906
Q4 1 8 9 Q2N3904
Q5 11 12 0 Q2N3904
.AC DEC 10 1 1MEG
.TRAN .05M 2.5M 0 .025M
.LIB
.OP
.PROBE
.END
```

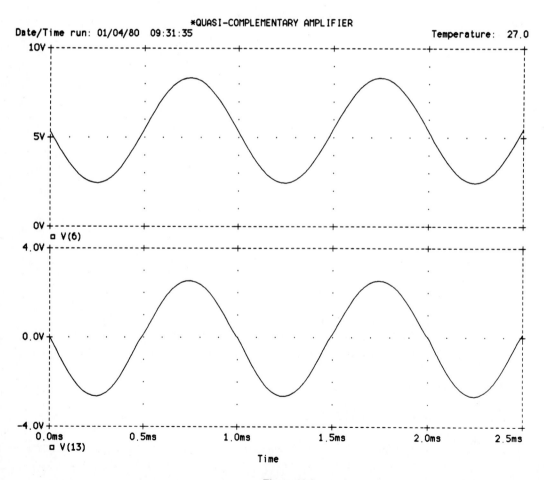

Figure 10-9

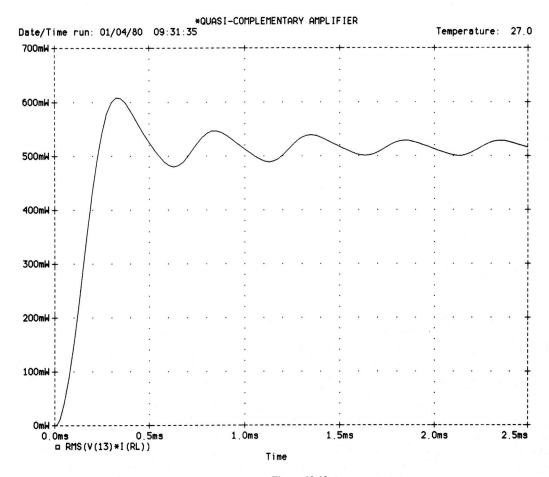

Figure 10-10

Note that the .OP command is used in the netlist; this allows us to see the operating conditions of all the transistors and diodes. The input impedance of this amplifier is moderate, in the range of 10 kΩ. The frequency range of this amplifier is also wide, greater than 100 kHz. The frequency range versus gain of the amplifier is shown in Figure 10-8. The voltage output at the collector of the input amplifier and the output at the load resistor are shown in Figure 10-9. The rms power of the amplifier is shown in Figure 10-10.

10.3. CLASS C RF POWER AMPLIFIERS

When working with RF, it is often necessary to match two widely different resistances, one in the collector of a transistor and the other, the load resistance of an antenna. To do this, several different types of network are used. If the collector resistance of the

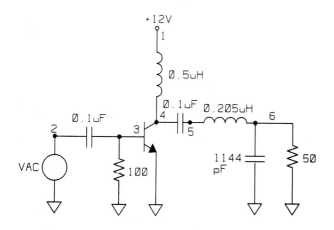

Figure 10-11

power amplifier is smaller than the load resistance, an *L* matching network can be used. If the collector resistance is higher than the load resistance, a pi network is often used. The conduction angle of the power amplifier is also important, as it determines the efficiency of the amplifier. The usual conduction angle of the amplifier is between 120° and 150°. PSpice can be used to analyze the amplifier for all the necessary parameters.

The first analysis we shall do will be for a transistor that produces 20 W of power into a 50-Ω load. The amplifier is operating at 10 MHz. The collector resistance of the power transistor is 3.6 Ω for this analysis. Since the collector resistance is less than the load resistance, we shall use an *L*-matching configuration with a coupling capacitor to keep dc out of the load. The transistor is a generic type of device with the following specifications.

```
.MODEL QMOD NPN(BF=150 IS=IE-8 CJC=15P CJE=30P)
```

The circuit of the amplifier is shown in Figure 10-11. The netlist of the amplifier is

```
*CLASS-C AMPLIFIER CHAPTER 10 PROGRAM 4
VCC 1 0 12
VAC 2 0 AC 1 SIN(0 12 10MEG)
CIN 2 3 .1U
RIN 3 0 100
LRFC 4 1 .5U
LTUN 5 6 .205U
CC 4 5 .1U
CTUN 6 0 1144P
RL 6 0 50
Q1 4 3 0 QMOD 1000
.MODEL QMOD NPN(BF=150 IS=1E-8 CJC=15P CJE=30P)
.TRAN .01U 2U
.PROBE
.END
```

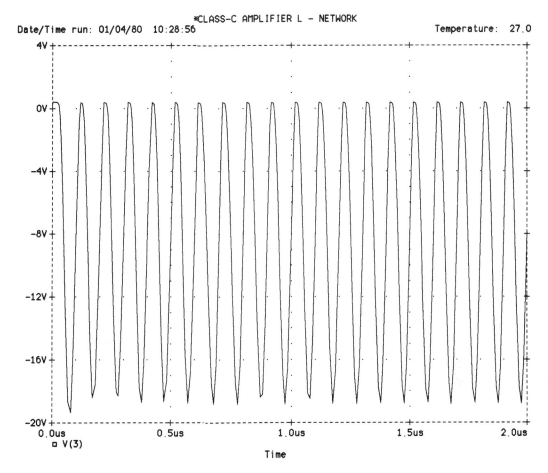

Figure 10-12

The analysis is shown in Figures 10-12, 10-13, and 10-14. From the analysis we find that the output of the amplifier is near 20 W within 2 μs.

The second of the two amplifiers we will analyze is a 0.25-W amplifier that couples its output to a 75-Ω load. The circuit of this amplifier is shown in Figure 10-15. The pi network is used in this amplifier. The pi network can be used with any combination of input and load impedance. It is especially useful when the output resistance of the driver device is larger than the load impedance. The netlist for the amplifier is

```
*0.25 WATT CLASS-C AMPLIFIER-CHAPTER 10
VCC 1 0 12
VAC 4 0 SIN(0 5 200K)
L1 1 2 1M
R1 3 0 100
```

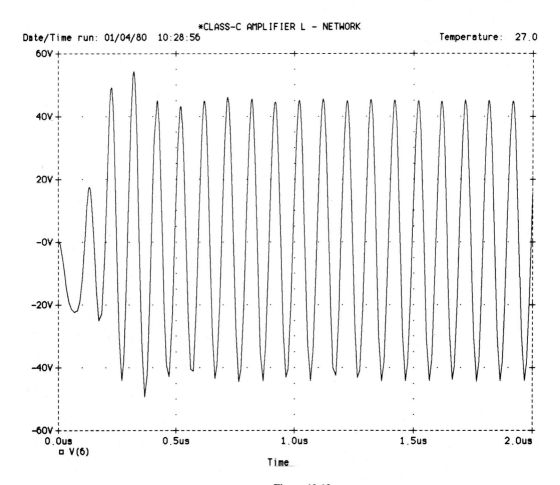

Figure 10-13

```
CC 4 3 .25U
CC1 2 5 .5U
C1 5 0 .041U
C2 6 0 .08U
L2 5 6 23U
RL 6 0 75
Q1 2 3 0 Q2N2222A
.LIB
.PROBE
.TRAN 5U 50U
.END
```

The analyses are shown in Figures 10-16, 10-17, and 10-18.

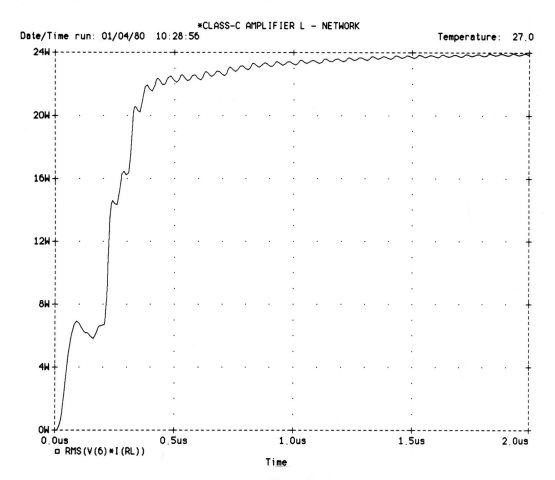

Figure 10-14

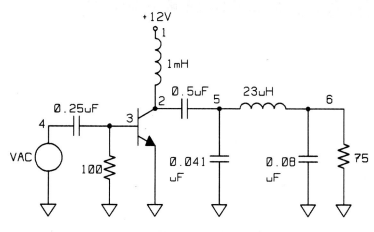

Figure 10-15

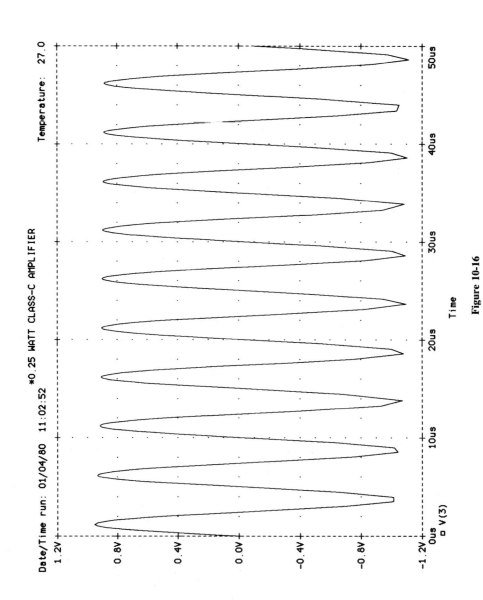

Figure 10-16

268

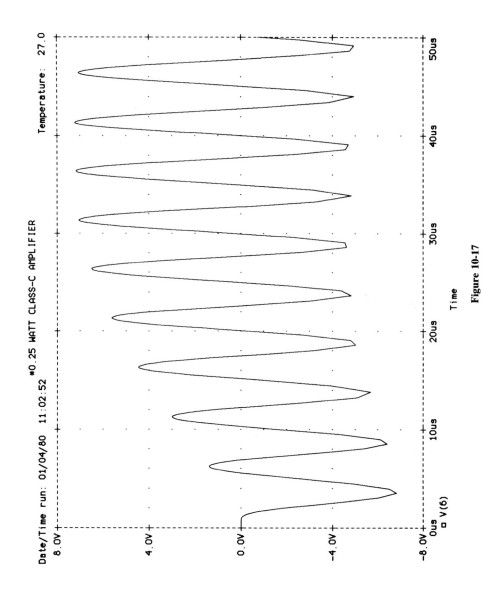

Date/Time run: 01/04/80   11:02:52   *0.25 WATT CLASS-C AMPLIFIER

Temperature: 27.0

Figure 10-17

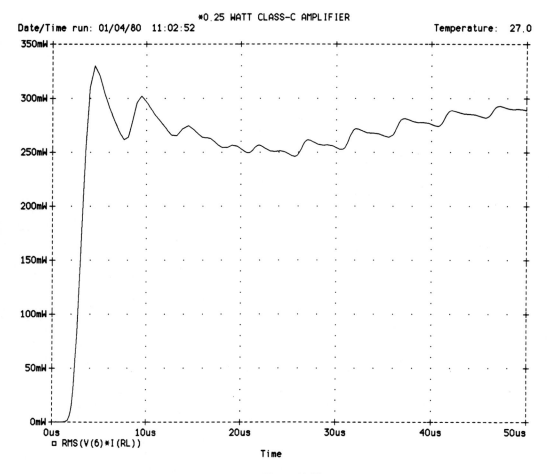

Figure 10-18

SUMMARY

PSpice can be used to analyze class A, B, and C power amplifiers. The power output of an amplifier into a load can be calculated using PROBE.

SELF-EVALUATION

1. The circuit shown in Figure P10-1 is a class B amplifier that uses an ideal op amp. Analyze this amplifier using both ac and transient analysis to determine both frequency response and distortion of the amplifier.

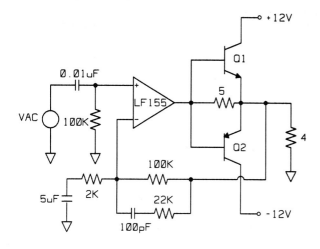

Figure P10-1

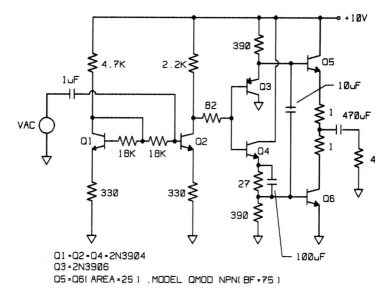

Q1 = Q2 = Q4 = 2N3904
Q3 = 2N3906
Q5 = Q6 (AREA = 25) .MODEL QMOD NPN(BF = 75)

Figure P10-2

2. The circuit shown in Figure P10-2 is a low-power class B amplifier for general-purpose use. Analyze this amplifier using both frequency response and transient analysis. With Fourier analysis, determine the distortion of the amplifier and try to modify it to provide less than 1% THD.

3. The circuit shown in Figure P10-3 is the power output stage proposed for a hand-held transmitter for 15 MHz. The power output is to be 2.5 minimum. Analyze the circuit and add any necessary components to achieve this power level. The transistor is to have the following parameters, and the area value for the transistor is 20.

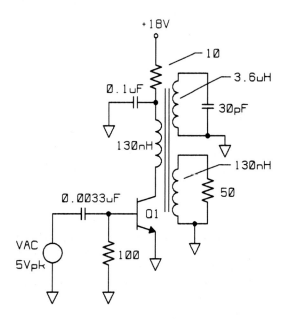

Figure P10-3

NPN(BF=150 IS=1E-14 RC=1 RE=.2 CJC=10P CJE=30P)

4. The circuit shown in Figure P10-4 is an amplifier that is used to show students the effects of class C bias on a JFET. Run an analysis of this circuit and determine if class C bias is in fact established. The power delivered to the 1000-Ω load should be about 10 mW when the circuit is operating properly.

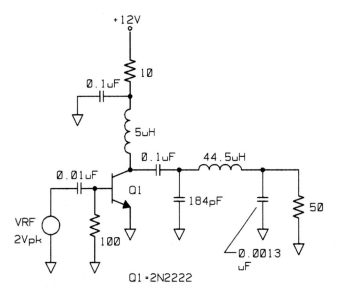

Q1 = 2N2222

Figure P10-4

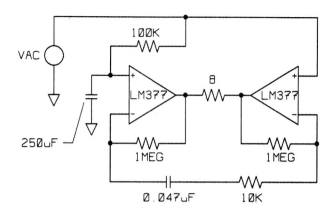

Figure P10-5

5. The circuit shown in Figure P10-5 is a bridge-type amplifier. This amplifier is made with two ideal amplifiers that simulate an LM377 integrated circuit power amplifier. As with all ideal amplifiers, this amplifier will produce all the power necessary in a circuit. For this reason there are two generic zener diodes across the output of the subcircuit. The diodes limit the excursion of the output voltage. The netlist for the subcircuit is

```
.SUBCKT LM377 NONINV INV OUT
RIN NONINV INV 3MEG
EGAIN A 0 POLY(2) (NONINV,0) (INV,0) 0 3.1623E4 -3.1623E4
ROL A B 1K
COL B 0 .0884U
EOUT C 0 B0 1
ROUT C OUT 25
D1 OUT D DMOD
D2 0 D DMOD
RD D 0 1E9
.MODEL DMOD D(BV=13)
.ENDS
```

Produce the following outputs for this amplifier:
a. The frequency response versus gain in decibels.
b. The clipping levels for an input signal of 1 V.
c. The distortion at 1 kHz.

Communication Circuits

OBJECTIVES

1. To learn to use PSpice to analyze common communications circuits.
2. To learn to use PSpice to analyze transmission lines.

INTRODUCTION

PSpice can be used at frequencies up to 1000 MHz or more. With this wide a frequency range it is only natural that we use it to analyze radio-frequency systems. You should also be aware that there are other versions of SPICE that are specially designed for the treatment of microwave and other specialized systems. In this chapter we analyze RF amplifiers, mixers, and detectors that are used in communication systems. Most of the analyses will be for discrete devices, but the techniques are equally applicable to integrated circuit communication systems.

11.1. RF AMPLIFIER ANALYSIS

There are two fundamental forms of RF amplifier in common use: the common-emitter (source) and common-base (gate) amplifier. We shall analyze each of the types.

Figure 11-1 shows a common-emitter amplifier that is designed to operate at 3.5

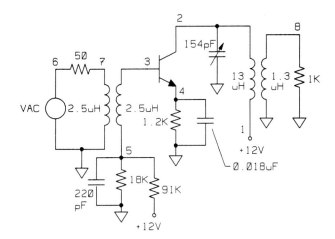

Figure 11-1

MHz. Note that some test points have been added to the circuit so that the ac voltages at the collector and base may be monitored. This is necessary since the ac voltages at these points are very small compared to the dc voltage. Also note that a magnetic model is used for the input transformer. This allows us to specify a turns ratio rather than an inductance for each winding. The model is an RF toroidal core with the following parameters:

```
.MODEL K59_43_000301 CORE(MS=1.9889E5 AREA=0.133 PATH=3.02
+        ALPHA=.00001 A=12 C=.001 K=18
```

The netlist for the amplifier is

```
*RF AMPLIFIER ANALYSIS
VDC 1 0 12
VAC 6 0 AC 1 SIN(0 1E-5 3.5MEG)
RB1 5 1 91K
RB2 5 0 18K
RE 4 0 1.2K
RS 6 7 50
CBP 5 0 220P
CE 4 0 .018U
L1 7 0 15
L2 3 5 15
KIN L1 L2 .9 K59_43_000301
CTEST 3 9 .1U
RTEST 9 0 1E7
CTST 2 10 .1U
RTST 10 0 1E7
CT 2 0 154P
```

```
CSTR 2 0 4P
LT 2 1 13U
LS 8 0 1.3U
RL 9 0 1K
KOUT LT LS 0.8
Q1 2 3 4 QMOD
.MODEL QMOD NPN(PE=0.6 CJC=2.5P CJE=12P VAF=50)
.LIB
.AC DEC 150 1MEG 10MEG
.TRAN .05U 4U
.PROBE
.OP
.FOUR 3.5E6 V(2) V(8)
.END
```

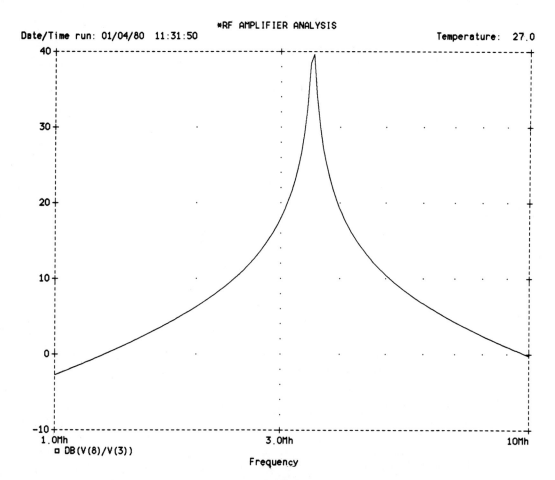

*RF AMPLIFIER ANALYSIS

Date/Time run: 01/04/80 11:31:50   Temperature: 27.0

□ DB(V(8)/V(3))

Frequency

Figure 11-2

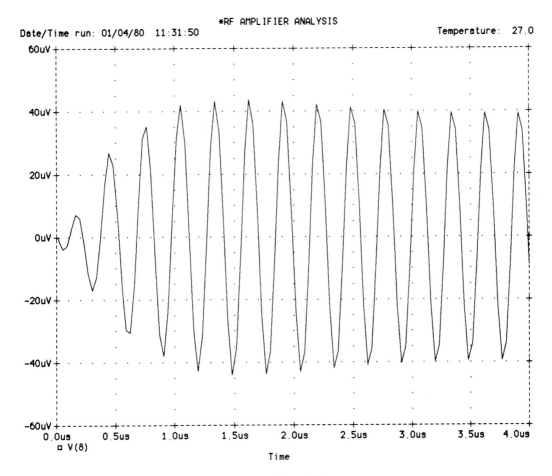

Figure 11-3

The parameters for the amplifier are as follows: f_{res} = 3.5 MHz, Q_L = 17.5, BW = 200 kHz, $C_{b'c}$ = 1 pF, $C_{b'e}$ = 2 pF, C_{stray} = 3 pF, A_v = 192, r_c = 10 kΩ. The output of the analysis is shown in Figures 11-2 and 11-3.

Since PSpice assumes a linear model for the ac analysis, we can have very large unrealistic voltages at the output points of the amplifier. To circumvent this, we use the decibel output of the amplifier since it will not change regardless of the input or output signal amplitude. This figure shows the amplitude of the output voltage of the amplifier for a more realistic signal input: in this case, 10 μV.

The second type of RF amplifier we will analyze uses a JFET with a tuned input and output. The amplifier is a common-gate type that operates at 50 MHz and is shown in Figure 11-4. This type of amplifier is popular at higher frequencies since it does not require neutralization. The gate effectively acts as a shield between the input and the

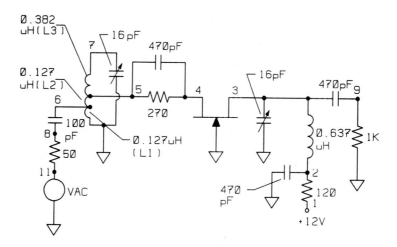

Figure 11-4

output. Also there is no Miller effect capacitance since the drain-to-gate capacitance is directly connected to ground. The netlist for the amplifier is

```
*COMMON GATE RF AMPLIFIER
VDC 1 0 12
VAC 11 0 AC 1 SIN(0 .01 50MEG)
RAC 11 8 50
R1 1 2 120
R2 5 4 270
RL 9 0 1K
C1 7 0 16P
C2 5 4 470P
C3 3 0 16P
C4 2 0 470P
CC 8 6 100P
CL 3 9 470P
L1 6 0 .127U
L2 6 5 .127U
L3 5 7 .382U
L4 2 3 .637U
J1 3 0 4 J2N3819
.LIB
.PROBE
.TRAN 1N 200N
.AC DEC 100 10MEG 1000MEG
.OP
.END
```

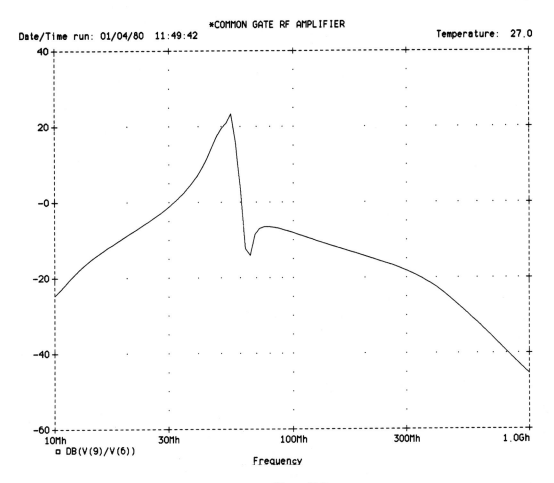

Figure 11-5

The output analysis of the amplifier is shown in Figures 11-5 and 11-6.

Another form of RF amplifier that is used to overcome the effects of Miller capacitance and the low input impedance of the common base circuit is the cascode amplifier. This amplifier requires two transistors in the common-emitter, common-base configuration shown in Figure 11-7. If the emitter currents of the two transistors is the same, the voltage gain of the common-emitter stage is then 1. Thus the Miller capacitance is minimized. The input impedance of the amplifier is that of the common-emitter stage. The gain of the circuit is the gain of the common-base amplifier portion of the cascode amplifier. The netlist for the amplifier is

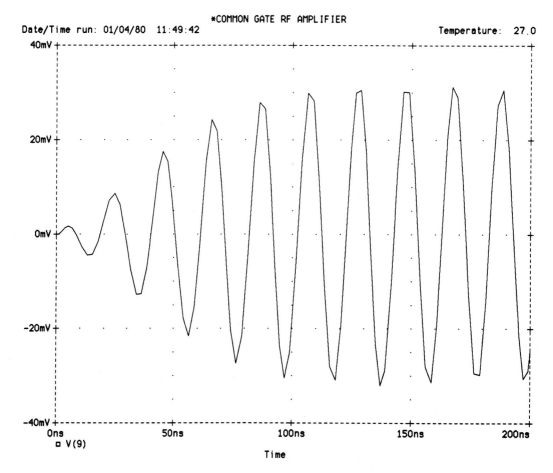

Figure 11-6

```
CASCODE RF AMP.
VDC 1 0 12
VAC 8 0 AC 1 SIN(0 .01 30MEG)
R1 6 0 8.2K
*R2 1 6 56K
R3 5 0 1K
R4 10 6 22K
R5 1 10 27K
C1 6 0 .001U
C2 5 0 .003U
C3 10 0 100P
CT 2 0 35P
```

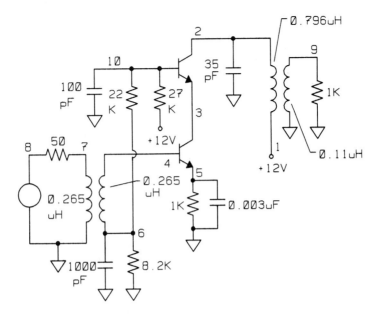

Figure 11-7

```
LT 1 2 0.796U
LS 9 0 0.11U
KT LT LS 0.9
RL 9 0 1K
RS 7 8 50
L1 7 0 .265U
L2 6 4 .265U
K2 L1 L2 0.9
Q1 2 10 3 QMOD
Q2 3 4 5 QMOD
.MODEL QMOD NPN(BF = 50 VAF = 50)
.PROBE
.AC DEC 200 1MEG 100MEG
.TRAN .03U 500N 0 1N
.END
```

The output of the analysis is shown in Figures 11-8 and 11-9.

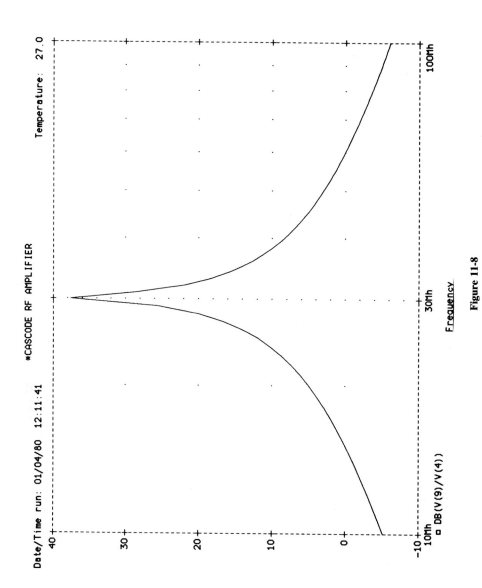

Figure 11-8

282

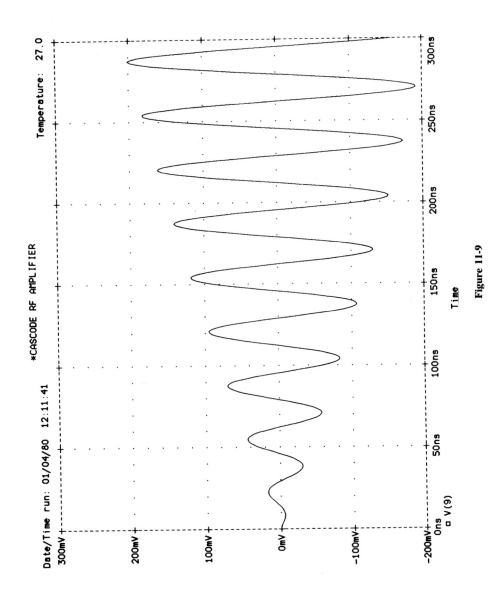

Figure 11-9

283

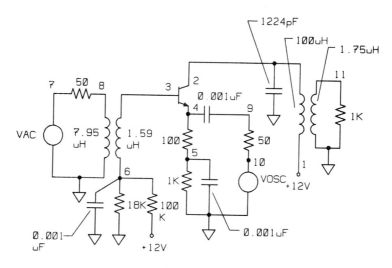

Figure 11-10

11.2. RF MIXERS

All radios of the superheterodyne type use mixers to convert the incoming RF signal to a lower or higher frequency. Whether the radio is in the form of an integrated circuit or discrete devices, the principles of mixing are the same. For mixing to occur, a non-linear device must be present. The nonlinear device is usually a transistor, but may also be a diode. We will examine several different types of mixer.

The first mixer is a transistor type in which the local oscillator signal is higher than the RF signal frequency. The circuit is shown in Figure 11-10. Note that we also do a Fourier analysis to determine the amount of distortion in the output 455-kHz signal. The netlist for the mixer is

```
AM MIXER
VDC 1 0 12
VRF 7 0 SIN(0 .01 5MEG)
VOSC 10 0 SIN(0 .5 5.455MEG)
VTEST C 0 AC 1
RTEST C D 100K
CTEST D 2 .001
.AC DEC 50 10K 1MEG
R1 6 0 18K
R2 6 1 100K
R3 5 0 1K
RS1 7 8 50
RS2 10 9 50
L1 8 0 7.95U
```

```
L2  3  6  1.59U
K12 L1 L2 0.8
COSC 9 4 .001U
RE  4  5  100
C1  6  0  .001U
C2  5  0  .001U
CT  2  1  1224P
LT  2  1  100U
LS  11 0  1.75U
KTS LT LS .9
RL  11 0  1K
Q1  2  3  4  QMOD
.MODEL QMOD NPN(BF=100 VAF=50 RE=1 RC=2)
.PROBE
```

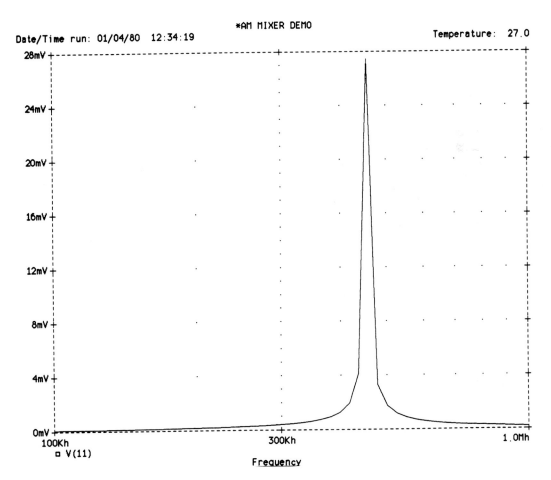

Figure 11-11

```
.TRAN .2U 20U
.OP
.FOUR 455K V(2) V(11)
.LIB
.END
```

The output of the mixer is shown in Figures 11-11, 11-12, and 11-13.

Let us now examine a sampling mixer. This type of mixer is excited by a square or rectangular waveform. The input waveform must be strong enough to switch the transistor completely on and off each cycle of the waveform. For our switching waveform we will use the pulse waveform, which is allowed to repeat for as many cycles as necessary. The number of cycles is determined by the .TRAN command maximum time limit.

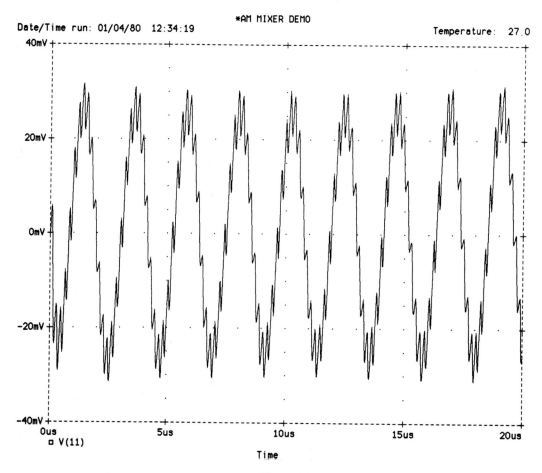

Figure 11-12

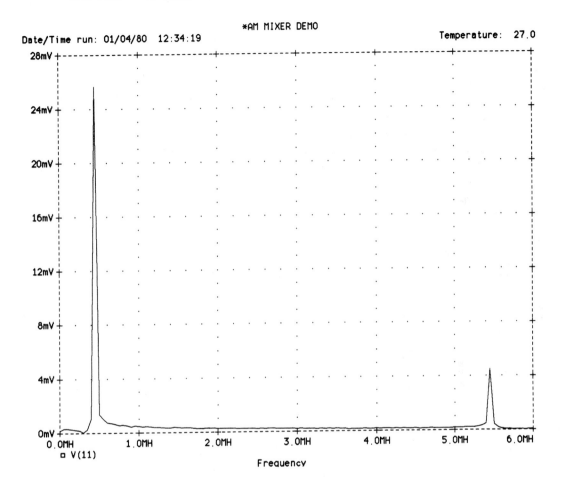

Figure 11-13

We shall use a simple circuit for this mixer; the only tuned circuit will be in the output of the transistor. The circuit of the mixer is shown in Figure 11-14. The netlist for the mixer is

```
*SAMPLING MIXER
VDC 1 0 20
VAC 6 0 AC 1 SIN(0 0.0005 5MEG)
VSAMPLE 8 0 PULSE(0 1 0 1N 1N 8.966E-8 1.833E-7)
RAC 6 A 300
R1 1 3 180K
R2 3 0 27K
R3 4 5 1K
R4 5 0 1K
```

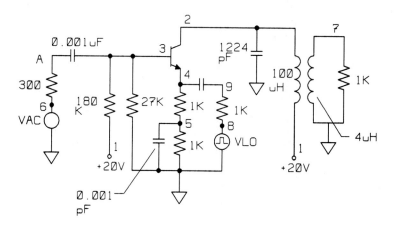

Figure 11-14

```
RSAMPLE 8 9 1K
C1 A 3 .001U
C2 5 0 .001U
C3 9 4 .001U
C4 2 1 1224P
L1 1 2 100U
L2 7 0 4U
K12 L1 L2 0.9
RL 7 0 1K
Q1 2 3 4 Q2N2222A ;QMOD
.MODEL QMOD NPN(BF=150)
.PROBE
.LIB
.TRAN 10N 20U
.AC DEC 50 100K 10MEG
.FOUR 455K V(7) V(2)
.OPTIONS ITL5=15000
.END
```

The .OPTIONS command is used in this analysis since the number of iterations for the solution will be greater than the default value of 5000. This analysis also generates a large amount of data. To run the full 20 μs, a hard disk system is a necessity. The outputs for this analysis are shown in Figures 11-15 and 11-16.

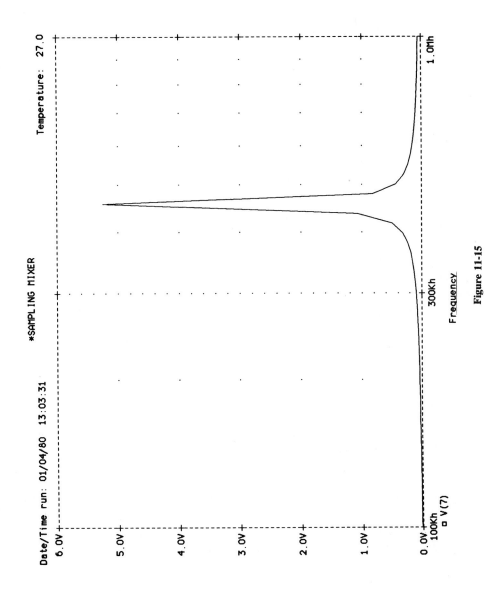

Date/Time run: 01/04/80 13:03:31   *SAMPLING MIXER   Temperature: 27.0

6.0V

5.0V

4.0V

3.0V

2.0V

1.0V

0.0V
100Kh   300Kh   1.0Mh
□ V(7)   Frequency

Figure 11-15

289

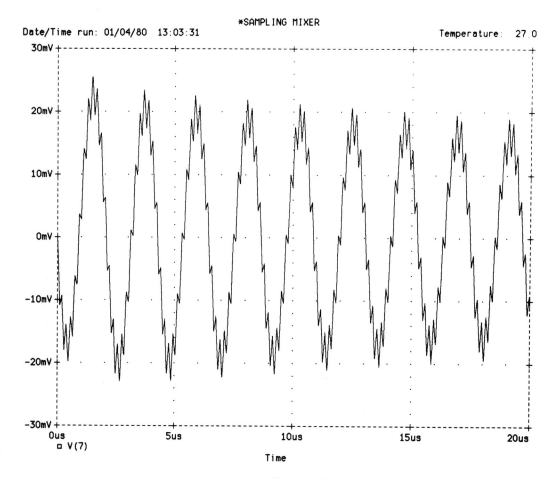

Figure 11-16

11.3. DETECTOR CIRCUITS

Of course, receiving and amplifying signals is important, but it is also necessary to extract the intelligence on the signal. To do this, we use a demodulator. The demodulator may be one of several different types. In the case of AM, a single half-wave rectifier that follows the peaks of the modulation is used. For FM or single sideband, other methods must be used. We shall analyze at least one each of several different types of detector circuit.

11.3.1. AM Detectors

The circuit for a simple AM detector is shown in Figure 11-17. The maximum depth of modulation for this detector is $m = 0.9$. The frequency of the IF is 455 kHz, and the diode is a germanium diode. The germanium diode is used for its low-voltage drop,

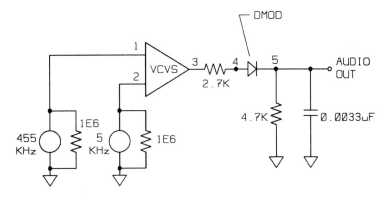

Figure 11-17

which requires a minimum carrier level of about 0.2 V to provide distortion free de-modulation. The maximum modulation frequency to be demodulated is 5 kHz. To use the detector, we must input a signal that is amplitude modulated to the desired depth. To do this we must first make a simple modulated source. The source must also reflect the proper impedance to the detector. For this analysis the detector input impedance will be about 2.2K Ω.

```
*AM DETECTOR
VCAR 1 0 SIN(0 2 455K)
R1 1 0 1E6
VMOD 2 0 SIN(1 .9 5K)
RMOD 2 0 1E6
EMOD 3 0 POLY(2) (1,0) (2,0) 0 0 0 0 1
RL1 3 4 2.7K
D1 4 5 DMOD
RD 5 0 4.7K
CD 5 0 .0033U
.MODEL DMOD D(IS=8.53E-6 RS=10)
.OPTIONS ITL5 = 10000)
.LIB
.PROBE
.TRAN 50U .4M
.AC DEC 25 100 100K
.END
```

Since we are calculating the parameters of two widely separated frequencies, the calculation time for this detector is long. To minimize calculation time, you should use the highest modulation frequency you expect to demodulate. The analysis will also use more than the 5000 iterations permitted by the ITL5 parameter. This limit is changed using the .OPTIONS command, as shown. Curves of the frequency response and de-modulated 5-kHz audio voltage are shown in Figure 11-18.

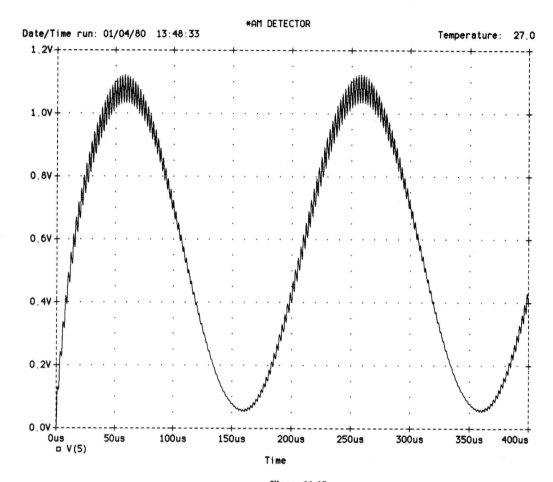

Figure 11-18

11.3.2. Product Detectors

Single-sideband transmission allows the conservation of spectrum space. It also provides an increase in the signal at the receiving antenna because the transmitted sideband can be amplified to the same power level the carrier is in double-sideband systems. However, it requires a more elaborate detection scheme than simple AM. Since there is no carrier transmitted with the sideband, the carrier must be reinserted at the receiver. There are several ways to do this. The one to be demonstrated here is the product detector. It is to be remembered that these analyses require long execution times and generate a large amount of data. The circuit of the product detector is shown in Figure 11-19.

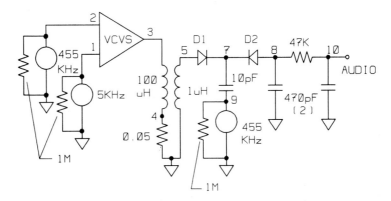

Figure 11-19

The DSB-SC generator is contained in the first few lines of the netlist. It is made from a VCVS source operated as a second-order polynomial device. If you remember, the equation for the second-order polynomial is

$$a_0 + a_1v_1 + a_2v_2 + a_3v_1^2 + a_4v_1v_2 + \cdots$$

The term of interest here is the last term, $a_4v_1v_2$. All the other coefficients are zero. The resistors in the VCVS are to provide a dc path to ground for each of the terminals. That this is a DSB-SC source can be seen by running just the VCVS device, commenting out the other components. Then examine the voltage at node 3 using the Fourier transform function supplied by PROBE. The netlist for the product detector is

```
* PRODUCT DETECTOR FOR AM SIGNALS
VCAR 9 0 SIN(0 5 455K)       ;SSB CARRIER
VAC1 1 0 SIN(0 2.25 5K)      ;SSB MODULATING SIGNAL
VAC2 2 0 SIN(0 2.25 455K)    ;REINSERTED CARRIER
EPOLY 3 0 POLY(2) (1,0)(2,0) 0 0 0 0 1 ;SSB GENERATOR
RPOLY1 1 0 1E6
RPOLY2 2 0 1E6
L1 3 4 100U
R1 4 0 .05
L2 5 0 1U
K12 11 12 0.9
D1 5 7 DMOD
D2 8 7 DMOD
CMOD 7 9 10P
RMOD 9 0 1E6
ROUT 8 10 47K
COUT 8 0 470P
C2OUT 10 0 470P
```

```
.MODEL DMOD D(IS=8.53E-6 RS=10)
.TRAN .01M .5M
.OPTIONS ITL5=0
.PROBE
.FOUR 5K V(10)
.END
```

The amplitudes of the carrier and modulation voltage were chosen to provide a large peak-to-peak voltage, such as would be found at the last IF amplifier of a radio. The level of the carrier should be at least several times that of the modulated waveform. The output of the product detector is tested using the .FOUR command. In this analysis the detected output has about 2% distortion. The output of this analysis is shown in Figures 11-20, 11-21, and 11-22.

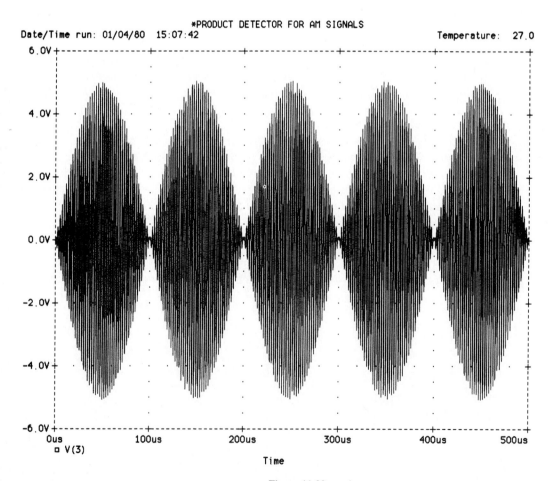

Figure 11-20

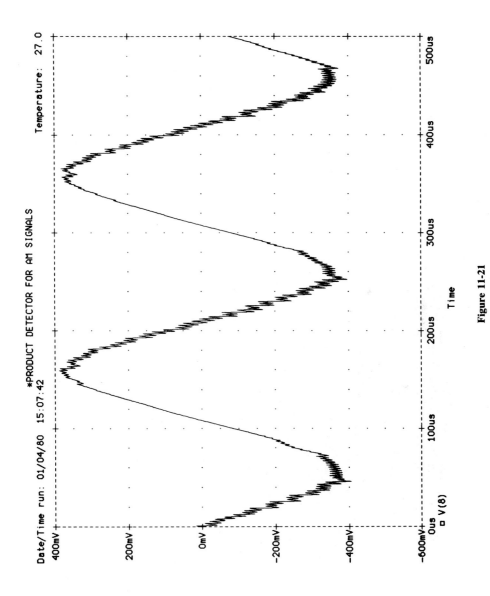

Figure 11-21

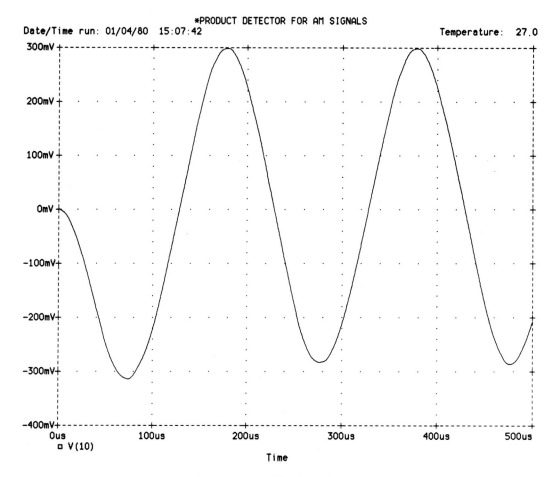

Figure 11-22

11.3.3. FM Detectors

There are several types of FM detectors. The most popular are the Foster–Stanley discriminator and the ratio detector. Of late, the PLL detector has also become very popular. The circuits of the Foster–Seeley discriminator and the ratio detector are shown in Figures 11-23 and 11-24.

Foster–Seeley discriminator   The Foster–Seeley discriminator uses a double-tuned circuit to differentiate an FM signal and produce an audio output. Note that this type of detector responds to AM modulation as well as FM modulation. For this reason a limiting amplifier is usually used as the last IF amplifier in the system. The analysis of this detector is best done using an ac sweep and then looking at the discrim-

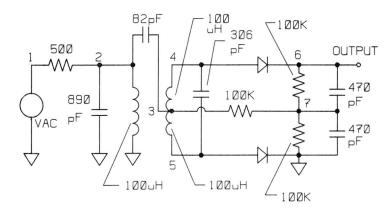

Figure 11-23

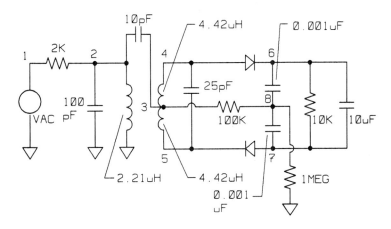

Figure 11-24

inator S curve. Both the input and output circuits are tuned to the center IF frequency. In this case the IF frequency is 455 kHz.

The coupling capacitor between the input and output coils is important because its reactance is reflected back to the primary, where it has a large effect on the resonant frequency of the tuned circuits. The diodes are a generic form of germanium diode. The netlist of this discriminator is

```
*FOSTER-SEELEY DISCRIMINATOR
VAC 1 0 AC 1
R1 1 2 500
C1 2 0 890P
L1 2 0 100U
CC 2 3 82P
```

```
L2 4 3 100U
L3 5 3 100U
K123 L1 L2 L3 0.999
C2 4 5 306P
RL4 3 7 100K
D1 4 6 DMOD
D2 5 0 DMOD
CF1 6 7 470P
CF2 7 0 470P
RF1 6 7 100K
RF2 7 0 100K
.MODEL DMOD D(IS=8.53E-6)
.AC LIN 300 100K 3MEG
.PROBE
.END
```

The output of the analysis showing the detector S curve is shown in Figure 11-25. Note that a linear sweep was used for this analysis.

Ratio detector   The analysis of the ratio detector is best done using just an ac sweep and then examining the S curve of the output voltage. It is not prudent to use an SFFM signal in this analysis. The modulating voltage will probably be in the audio range and to provide even one or two cycles of audio requires millisecond analysis times. For this reason, the analysis can take a long time and generate very large amounts of data.

The ratio detector S curve is observed as the imaginary part of the output voltage. Components can then be "tweeked" to provide the best detector linearity and band-width. The ratio detector is operating at 10.7 MHz and the diodes used are generic diodes of the germanium type. The netlist of a ratio detector is

```
*RATIO DETECTOR FOR FM
VAC 1 0 AC 1
R1 1 2 2K
C1 2 0 100P
L1 2 0 2.21U
L2 4 3 4.42U
L3 5 3 4.42U
K123 L1 L2 L3 0.9
CC 2 3 10P
CT 4 5 25P
D1 4 6 DMOD
D2 7 5 DMOD
CF1 6 8 .001U
CF2 8 7 .001U
RF1 6 7 10K
RF2 3 8 100K
```

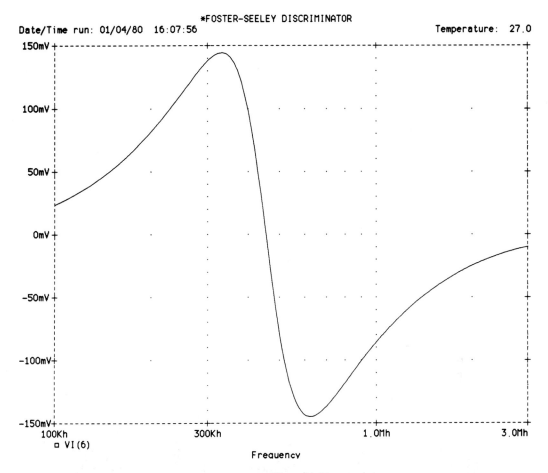

Figure 11-25

```
CAGC 6 7 10U
RL 8 0 1E6
.MODEL DMOD D(IS=8.53E-6)
.AC LIN 300 7MEG 15MEG
.PROBE
.END
```

The output S curve for this detector is shown in Figure 11-26. Note that we used a linear sweep for this analysis also.

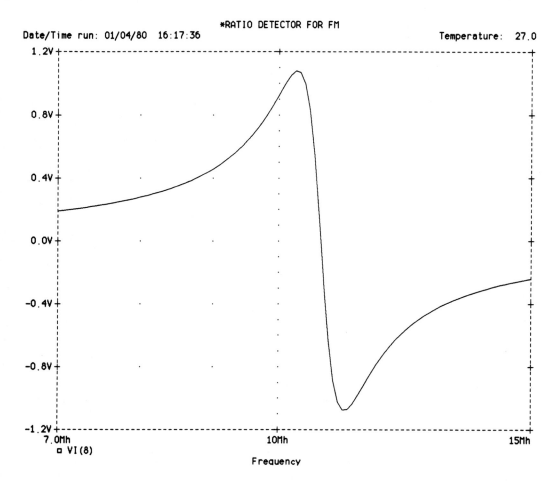

Figure 11-26

11.4. TRANSMISSION LINES

Transmission lines are used a number of ways in electronics. They may be used only to move electrical energy from one point to another, or they may be used as matching transformers for antennas, transmitters, and so on. In many cases the transmission is either an appreciable part of a wavelength at the frequency being used, or it may be a few to many wavelengths long. In any case, it is not just a piece of wire. It has special, predictable characteristics. As we learned in school, transmission lines are lossless, except when we get into the real world. In PSpice, transmission lines are bidirectional and truly lossless.

The general syntax for the transmission line is

```
T<name>  <+ node 1>  <- node 1>  <+ node 2>  <- node 2>
+    ZO = <value> [TD = <value>] [F = <value> [NL = <value>]]
```

The transmission line in PSpice is modeled as a two-port device. The positive nodes denote positive voltages at the ports. There is no provision for setting the L and C parameters for the line. We can, however, set both the frequency and wavelength of the line or its delay time. The parameter ZO is the characteristic impedance of the transmission line and must be specified. The length of the line is specified in one of two ways. You can either specify the time delay of the line, TD, or the frequency of operation, F, and the length of the line. At least one of these two properties must be specified. If NL is not specified with F, NL defaults to 0.25 wavelength. For the transmission line you may use either Z and the number zero, or Z and the letter O. Both are accepted by PSpice. During treatment analysis, the internal time step of PSpice is limited to one-half the smallest delay time. Because of this, short transmission lines cause long run times.

Some useful conversions for transmission lines

Time delay:

$$TD = \frac{L}{v}$$

where L is the physical length of the line and v is the phase velocity factor of the line.
   Velocity:

$$v = v_o \times \mathrm{vf}$$

where v_o is the free-space velocity and vf is the velocity factor as given by the manufacturer's literature.
   Normalized length: The normalized length of the line NL can be expressed as

$$NL = \frac{L}{\lambda}$$

where λ is the wavelength in the transmission line.
   Wavelength of the line: The wavelength in the transmission line is determined by

$$\lambda = \frac{v}{f}$$

Normalized length: The normalized length can then be expressed as

$$NL = \frac{Lf}{v}$$

   Time delay: Finally, the time delay, TD, and the normalized length can also be expressed as

$$TD = \frac{NL}{f}$$

$$NL = f \times TD$$

11.5. THE TRANSMISSION LINE AS A DELAY ELEMENT

The first analysis is for the transmission line as a delay line. We will do this for the case of a matched line and the line with a resistive termination of 0.5ZO and 2ZO. As you will see, PSpice can be used to set up the approximate display of a time-domain reflectometer (TDR). The effects of L and C added at the load end could also be tested for, and the resulting TDR display examined. The transmission line netlist is

```
*TRANSMISSION LINE 1 - ZO MATCHED.
VAC 1 0 AC 1 PWL(0,0 100P,10 300P,10 301P,0 4E-9,0)
RO 1 2 50
T1 2 0 3 0 ZO=50 TD=1.732N
RL 3 0 50
.TRAN .1N 4E-9 0 .01N
.PROBE
.END
*TRANSMISSION LINE 2 - 0.5ZO TERMINATION.
VAC 1 0 AC 1 PWL(0,0 100P,10 300P,10 301P,0 4E -9,0)
RO 1 2 50
T1 2 0 3 0 ZO=50 TE=1.732N
RL 3 0 25
.TRAN .1N 4E-9 0 .01N
.PROBE
.END
*TRANSMISSION LINE 3 - 2ZO TERMINATION.
VAC 1 0 AC 1 PWL(0,0 100P,10 300P,10 301P,0 4E-9,0)
RO 1 2 50
T1 2 0 3 0 ZO=50 TD=1.732N
RL 3 0 100
.TRAN .1N 4E-9 0 .01N
.PROBE
.END
```

The output of the analysis is shown in Figures 11-27, 11-28, and 11-29. You can see that for a matched line, the reflected pulse is at the delay time specified for the transmission line in the matched condition. For the mismatched condition the reflected pulse is at twice the delay time and is either positive or negative, depending on whether the mismatch is greater than or less than the ZO of the line.

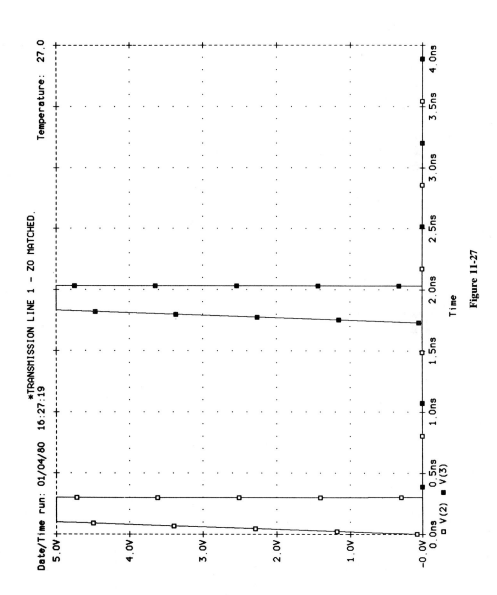

Figure 11-27

303

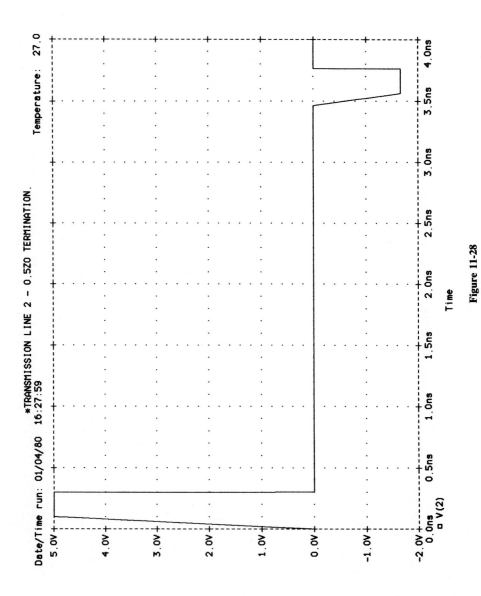

Date/Time run: 01/04/80   16:27:59   *TRANSMISSION LINE 2 - 0.5Z0 TERMINATION.   Temperature: 27.0

Figure 11-28

304

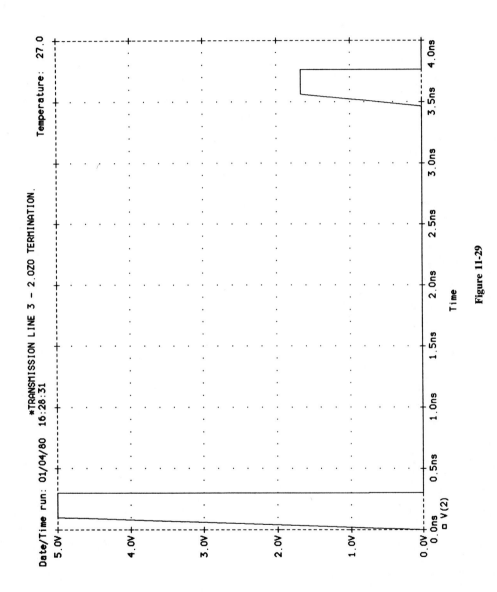

Figure 11-29

305

11.6. USING A TRANSMISSION LINE TO MATCH IMPEDANCES

One of the most common uses of the transmission line is to match an unbalanced transmission line (coaxial cable) to a balanced transmission line or antenna. Usually, the antenna being matched is a higher impedance than the coaxial cable being used for the transmission line. The transmission line can be used to effect a 4:1 impedance change if the line is $\frac{1}{2}\lambda$ long at the frequency of interest. Thus a 75-Ω coaxial cable can be matched to a 300-Ω antenna with a half-wave section of line. For this analysis the transmission line will be swept twice. The first sweep will allow us to determine the frequencies at which the response of the balun transformer is within -3 dB of the maximum output. We will set the frequency of the transmission line to 250 MHz for both of these analyses.

The second sweep allows us to see the input impedance by replacing the voltage source and source resistance with a current source. If you recall, the impedance of a device can be seen by allowing a current source to have 1 A. When this occurs, since $E = IZ$, the impedance is equal to the voltage because $I = 1$. There are also real and imaginary voltages on the transmission line. We use these to determine the SWR of the balun transformer. The SWR is useful to define the useful range of frequencies for the transformer.

The netlist for the balun transformer is

```
*4:1 BALUN TRANSFORMER
VRF 1 0 AC 1
RO 1 2 75
T1 2 0 3 0 Z0=75 F=250MEG NL=100
T2 3 0 4 0 Z0=75 F=250MEG NL=0.5
RANT 3 4 300
.AC LIN 100 50MEG 500MEG
.PROBE
.END
*4:1 BALUN MATCHING TRANSFORMER - IMPEDANCE
IAC 0 2 AC 1
T1 2 0 3 0 Z0=75 F=250MEG NL=1
T2 3 0 4 0 Z0=75 F=250MEG NL=0.5
RANT 3 4 300
.AC LIN 100 150MEG 350MEG
.PROBE
.END
```

The output analyses are shown in Figures 11-30 and 11-31.

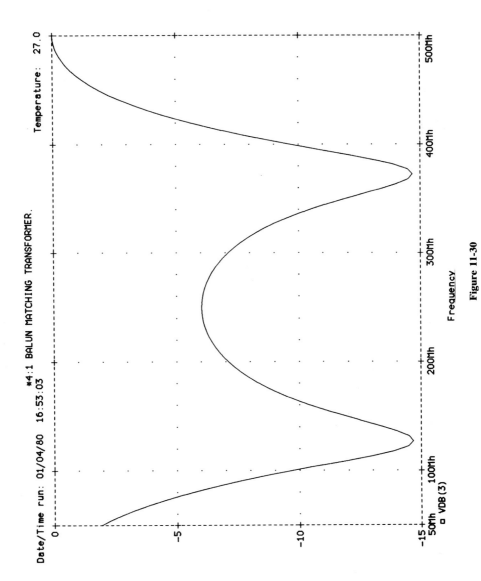

Date/Time run: 01/04/80 16:53:03   *4:1 BALUN MATCHING TRANSFORMER.   Temperature: 27.0

Frequency

Figure 11-30

307

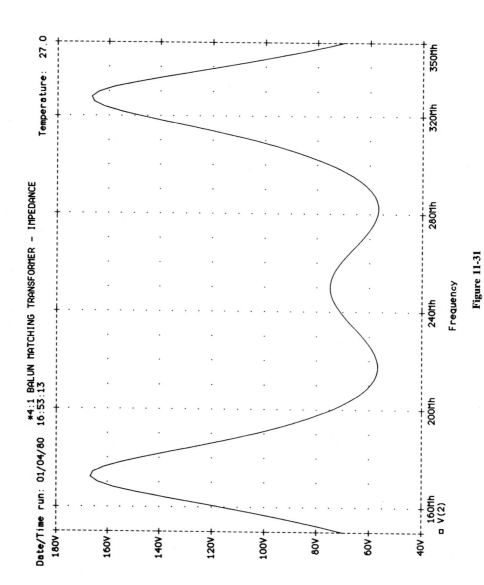

Date/Time run: 01/04/80   *4:1 BALUN MATCHING TRANSFORMER - IMPEDANCE   Temperature: 27.0
16:53:13

Figure 11-31

308

SUMMARY

PSpice can be used to calculate the parameters of RF amplifiers, mixers, and oscillators in subsystems. PSpice can also be used to calculate the output of various types of detector, AM detectors, the product detector, and the FM types included. PSpice controlled sources can be used to generate the required DSSB signals for the product type of detector. PSpice can be used to analyze transmission lines. The use of a transmission line in delay, and TDR applications is also possible.

SELF-EVALUATION

1. For the circuit of Figure P11-1, design and analyze the amplifier for use as a IF amplifier in a 10.7-MHz IF strip. The dc conditions for the amplifier are $I_e = 1$ mA, $V_{cc} = 12$ V, $V_e = 1.2$. The power gain of the amplifier is to be 26 dB, and its bandwidth is 400 kHz. The transistor used has the following model:

```
.MODEL QMOD NPN(IS=1E-18 BF=95 RC=1 RE=.5 CJC=.1P CJE=5P)
```

You should produce curves for the following:
a. Frequency response.
b. Output–input voltage gain.
c. Phase angle at the collector.

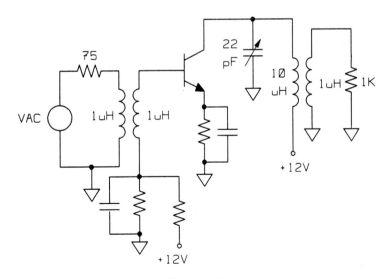

Figure P11-1

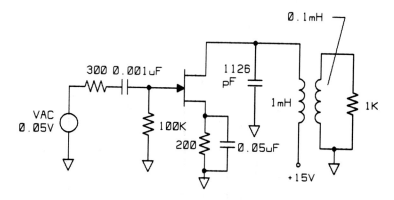

Figure P11-2

2. For the circuit of Figure P11-2, the JFET RF amplifier operates at 150 kHz. Produce operating curves for
   a. The output–input power gain.
   b. The phase shift at the drain.
3. The circuit of Figure P11-3 is a JFET mixer. Analyze and improve on the circuit.
4. Design the mixer shown in Figure P11-4 for use in an FM receiver with an IF frequency of 10.7 MHz. The RF frequency is 98 MHz. Let the input transformer have a 1:1 turns ratio and be 1 μH for each winding. The maximum IF power output is to be 4 mW, and the bandwidth is to be 500 kHz.
5. Design a transmission-line balun transformer to match 79-Ω antenna to a 300-Ω receiver at a frequency of 625 MHz.

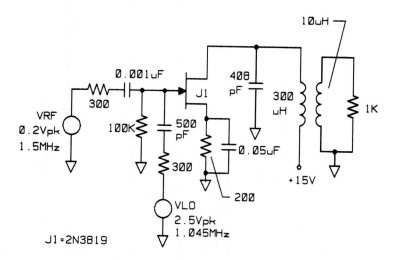

Figure P11-3

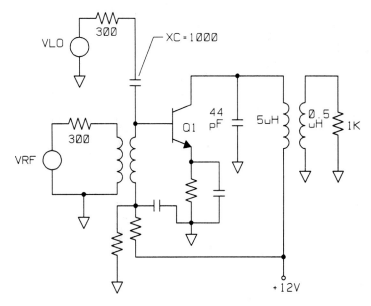

Figure P11-4

12

Additional PSpice Commands and Analysis Types

OBJECTIVES

1. To learn the use of PSpice for Monte Carlo types of analysis for determining the range of variation that can be expected in manufacturing products.
2. To show some of the other dimensions of PSpice that are not available in the student version.

INTRODUCTION

In this chapter we look at additional analyses and commands that PSpice can do. Some of the analyses shown in this chapter are not available in the student version of PSpice. The analysis types have been saved for last because they generate very large amounts of data. Let us start with the two analysis types to be discussed here, then go on to the additional commands.

12.1. .MC, THE MONTE CARLO ANALYSIS

In my first job as an engineer, after attending college, I was given the job of designing an audio amplifier for production. At that time I was told by my boss: "anyone can build one of anything." It was my job to make this amplifier producible in terms of thousands of them. When you must produce thousands of a given product, you are inevitably going to run into the limits of the tolerances of the devices that you specify. For this reason, you must determine the limits of performance of the product when devices that you have specified combine to produce the least acceptable performance and the maximum possible performance.

Monte Carlo analysis provides the vehicle to do a statistical analysis based on component tolerances that you supply. There are two built-in types of analysis. One uses a uniform distribution and the other a Gaussian distribution. The uniform distribution is the default analysis. In all cases, a run of the netlist using nominal values is done first; then the Monte Carlo analysis takes effect. There are three components to any Monte Carlo analysis:

1. Element tolerances
2. Multiple runs
3. Output

To perform a Monte Carlo analysis you must specify all three of these. The general syntax for the Monte Carlo analysis is

```
.MC <numruns> <analysis> <output> <function> [<option>]
```

where ⟨numruns⟩ is the number of runs the analysis is to make. ⟨analysis⟩ is the type of analysis performed, dc, ac, or transient. ⟨output⟩ is the output variable that will be tested. ⟨function⟩ is the operation performed on the ⟨output⟩, where ⟨function⟩ must be one of the following:

1. YMAX   the greatest difference in each waveform from the nominal value.
2. MAX   maximum value of each waveform.
3. MIN   minimum value of each waveform.
4. RISE_EDGE ⟨value⟩   first occurrence of the waveform above a threshold value.
5. FALL_EDGE ⟨value⟩   first occurrence of the waveform below a threshold value.
6. LIST   prints model parameter values used for the components during each run.
7. OUTPUT (output type)   where (output type) is data from runs subsequent to the first run, and must be one of the following types:
   a. ALL   generates all output from all runs.
   b. FIRST ⟨value⟩   output from the first ⟨value⟩ runs.
   c. EVERY ⟨value⟩   output every ⟨value⟩ runs.
   d. RUNS ⟨value⟩   generates output only for specified runs; up to 25 values may be listed.
   e. RANGE (⟨lowval⟩, ⟨highval⟩)   sweeps only parameters that are in the range of computation of ⟨function⟩. If the range is omitted, the total range of values is computed.

12.1.1. Tolerances for Monte Carlo Analysis

To do a Monte Carlo analysis, device parameters must be allowed to change. There are two ways to do this in PSpice. One is deviation of the device parameters only, the other is deviation of the lot device parameters. The tolerances are used inside the .MODEL statement. There are two keywords associated with tolerances: DEV and LOT. Each of these follows the parameter to be varied by the analysis. Also, each applies only to the parameter they follow. There is no way to vary all parameters of a device at the same time.

An example of DEV tolerance used within a .MODEL statement is

```
.MODEL QMOD PNP(BF=100 DEV=25% IS=1E-18 CJC=1P)
```

The forward beta of the transistor is the only parameter affected in this statement. It is important that you do not make the tolerance of the parameter larger than the parameter's value. If you do, the program will abort and no data will be generated. Note that the variation permitted is a percentage of the parameter's nominal value. If the percent sign were left off, the tolerance is an absolute value. If more than one device addresses this model, each of the devices will probably have different values for the same run of the analysis. Thus one transistor in an amplifier may have a beta of 120, while another may have a value of 90 within the same run. This is useful when the devices have uncorrelated tolerances.

An example of the line above with the tolerance changed to a LOT-type tolerance is

```
.MODEL QMOD PNP(BF=100 LOT=50% IS=1E-18 CJC=1P)
```

In this case the devices will have the same beta but will vary within the limits set by LOT. The first Monte Carlo run may have a beta value of 122 for all transistors. The next run may have beta equal to 88, once again for all transistors.

The DEV and LOT tolerances may be used together when the device parameters are not completely independent of one another. If some correlation is expected from such devices as a matched pair from the same production run, it may be best to use both tolerances. An example of both tolerances used on a device is

```
.MODEL QMOD PNP(BF=100 DEV=25% LOT=25% IS=1E-18 CJC=1P)
```

In this case, device tolerances could be as much as 50% different from their nominal value, but the devices cannot be more than 25% different from each other. Thus device and lot tolerances add. There is more to the Monte Carlo analysis than presented here. You should consult the MicroSim handbook for PSpice (MicroSim Corporation, 1989) for all the parameters possible with this analysis type.

12.1.2. **Running the Monte Carlo Analysis**

To get good data on the maximum and minimum limits of the component or system you are designing, it is necessary to do many Monte Carlo analyses. The more you do, the better the statistical data. However, the more runs you make, the longer it takes the computer to complete the analysis. In this case, the time taken is directly proportional to the number of runs made. If you do 50 runs, it takes 50 times as long as a single run. You must decide how much time, and how many runs is enough.

When the Monte Carlo analysis is done, there are three types of data available:

1. Model parameters with tolerances applied; this is available in the .OUT file.
2. Waveforms from each run; this is available using .PROBE, .PRINT, and .PLOT.
3. Summary of all runs using a collating function; this is available in the .OUT file.

Of course, the best way to understand the Monte Carlo analysis is to do one. Let us analyze the class A amplifier we analyzed in Chapter 10. We will run five Monte Carlo analyses, varying only the device forward beta. The netlist for the analysis is

```
*CLASS A AMPLIFIER - MONTE CARLO ANALYSIS
VCC 1 0 15
VAC 2 0 AC .1 SIN(0 .1 1K)
R1 1 4 12K
R2 4 0 2.2K
R3 5 0 50
RL 6 0 8
RG 2 12 600
RD 3 13 10
C1 7 4 10U
C2 5 0 4700U
*CL 3 0 .001U
L1 1 13 1
L2 6 0 .00444
K12 L1 L2 .9999
R4 1 7 7.5K
R5 9 0 1.5K
R6 1 8 120K
R7 8 0 22K
CIN 12 8 10U
Q1 3 4 5 Q2N3904
Q2 7 8 9 Q2N3904
.AC DEC 20 1 1MEG
.TRAN .01M .005 0 .025M
.MC 15 TRAN V(3) YMAX LIST OUTPUT ALL
.PROBE
.OP
.LIB
.OPTIONS NOPAGE
.END
```

Before you run this netlist, be sure that you have modified the 2n3904 model in the library. Your modified library file should look like this:

```
.MODEL Q2N3904 NPN(Is=6.734f Xti=3 Eg=1.11 Vaf=74.03
+               Bf=416.4 DEV=25% LOT=25% Ne=1.259
Ise=6.734f
+               Ikf=66.78m Xtb=1.5 Br=.7371 Nc=2 Isc=0
Ikr=0 Rc=1
+               Cjc=3.638p Mjc=.3085 Vjc=.75 Fc=.5
Cje=4.493p
+               Mje=.2593 Vje=.75+ Tr=239.5n Tf=301.2p
Itf=.4
+               Vtf=4 Xtf=2 Rb=10)
```

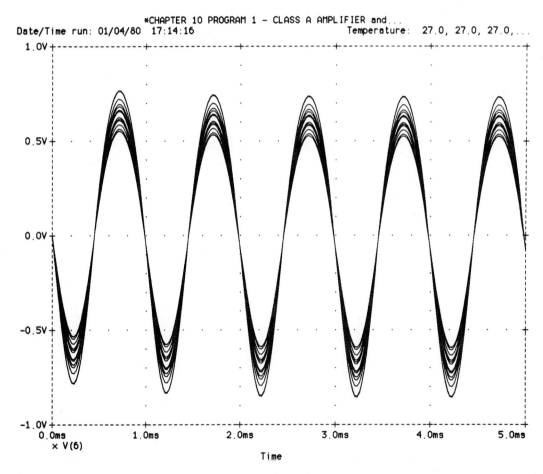

*CHAPTER 10 PROGRAM 1 - CLASS A AMPLIFIER and...
Date/Time run: 01/04/80 17:14:16   Temperature: 27.0, 27.0, 27.0,...

Figure 12-1

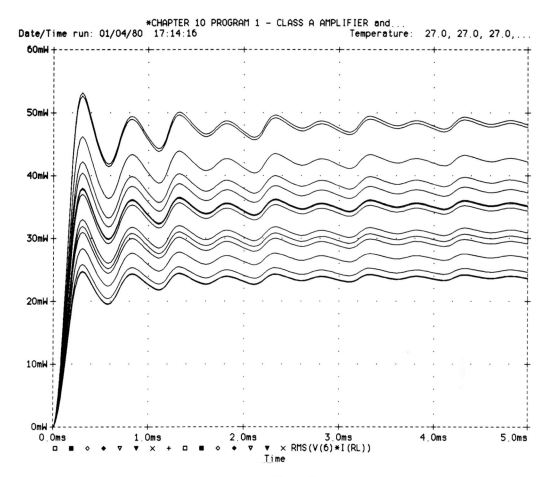

Figure 12-2

The device and lot tolerance will affect only the forward beta of the transistors. Both will be varied, but will not necessarily be the same value.

When the analysis is done, the first analysis is a run using the nominal values for the devices. The runs following this will be at different values of forward beta. The output of the analysis is shown in Figures 12-1 and 12-2 for the output voltage at node 6 and the rms power in the load resistor. The input voltage was changed from 0.1 V for the transient response to 0.13 V to bring us to the verge of clipping. Note that the power ranges from 40 mW to about 70 mW for all of the runs. If we had advertised a minimum output power of 50 mW maximum for this input voltage, we would have to do some redesign to meet this specification.

The output file is large for this analysis (about 75K BYTES). This is because we have caused the Monte Carlo analysis to generate almost all the data it can for this amplifier. Two of the outputs are reproduced in Figures 12-1 and 12-2. These outputs represent the worst case calculated for distortion at the output. You can see that for this amplifier distortion ranges from about 7.9% to about 11.33%.

While this analysis by no means covers all possibilities, it is probably adequate to indicate that some redesign is indicated. If you were really going to design something for production, you may wish to use many more runs than 15. In fact, you may have enough to allow you to go to lunch while the Monte Carlo is running.

12.2. ANALOG BEHAVIORAL MODELING *

Analog behavioral modeling is a useful analysis tool provided with PSpice. It allows us to write the description of components in terms of transfer functions. The transfer function can be an equation, a table, or in the form of a Laplace frequency-domain function. This analysis in PSpice allows us to use two of the controlled (dependent) sources as transfer function devices. The two devices used are the E, or VCVS device, which gives a voltage output, and the G, or VCCS device, which gives a current output. The F and H device cannot be used for behavioral modeling.

The rules for use of the controlled devices change significantly when they are used in this application. We shall not go into all of the possible uses of this function here. We concentrate on the requirements to use the function successfully to produce desired results.

12.2.1. Voltage and Current Sources

We begin by looking at the syntax sources that have specified outputs that can be expressed as a formula, or equation.

```
E<name> <+node> <-node> VALUE = {<expression>}
G<name> <+node> <-node> VALUE = {<expression>}
```

These are two forms of source that require no input nodes. Their output is {<expression>}. The equation you use may contain any mixture of voltages, currents, time, or other parameter. The primary concern in the use of these sources is that {<expression>} must fit on one line. You cannot use the plus (+) sign to extend the line. You can use many of the arithmetic operators that you use with PROBE, within the {<expression>}. Unfortunately, not all of the operators are available. A list of those operators that you may not use is given below.

```
SGN(X)
DB(X)
D(X)
S(X)
AVG(X)
RMS(X)
```

*MICROSIM PSpice Manual

An example of the E and G devices is

```
EFUNCTION 25 0 VALUE = {5*PWR(TIME,2)}
GDEVICE 6 0 VALUE = {I(V(5))-I(V(INPUT)}
```

As usual, the E device has a voltage output and the G device has a current output, just as they did in the controlled devices we studied. The word VALUE must have a space between it and the equal sign, and the braces around the expression are mandatory. The output voltage of this E device is five times the square of the time at which the analysis points are calculated. The voltage appears at nodes 25 and 0.

The output current of the G device is the current in the voltage source V(5) minus the current in the voltage generator V(INPUT). Note that the current must be measured through a voltage source. This is a case where a voltage source of zero volts can be used. A voltage source of zero volts, often called a *dead source,* is used for measuring current.

12.2.2. Voltage and Current Sources Using Tables

One way we may use the voltage and current source is as a device whose transfer function can be described by a table of data. It should be made clear that while useful, this can also become tedious. If the transfer function has many points, we can have many lines describing it in our netlist. The syntax for using the table form of voltage and current source is shown below.

```
E<name> <+node> <-node> TABLE {<expression>} =
+        <x parameter, y parameter> · · ·
G<name> <+node> <-node> TABLE {<expression>} =
+        <x parameter, y parameter> · · ·
```

For this type of device, the expression must fit on one line, and you cannot use the plus (+) sign to extend the line. The x and y parameters are the input–output data that are needed for the device. There may be as many data lines as needed. The data must be arranged in a lowest value to highest value form for the x parameters. You may have as many pairs of data as you need for the table. The plus (+) sign can be used to extend the line in this part of the source. An example of using this form of the sources is to provide a voltage and current source that output a parabolic voltage and current of the form $y = x^2$.

```
EPARABOLA 2 0 TABLE {V(20)} =
+ (0,0) (1,1) (2,4) (3,9) (4,16) (5,25) (6,36) (7,49)
+ (8,64) (9,81) (10,100)

GPARABOLA 2 0 TABLE {V(20)} =
+ (0,0) (1,1) (2,4) (3,9) (4,16) (5,25) (6,36) (7,49)
+ (8,64) (9,81) (10,100)
```

This source requires an input voltage at node 20, {V(20)}. The input voltage is a piece-wise linear voltage starting at 0 V and ending in 1 s at 10 V. The dependent source produces the parabolic output from the transfer function. A netlist for this source that produces an output that is parabolic is

```
*TABLE FUNCTIONS
VIN 20 0 PWL (0,0 1,10)
RIN 20 0 1E6
EPARABOLA 2 0 TABLE {V(20)} =
+ (0,0) (1,1) (2,4) (3,9) (4,16) (5,25) (6,36) (7,49)
+ (8,64) (9,81) (10,100)
ROUT 2 0 1E6
```

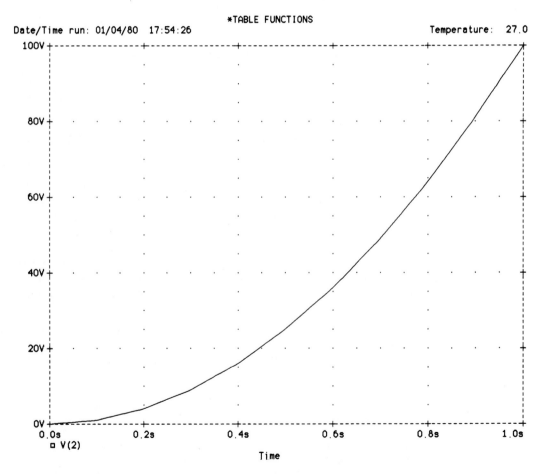

*TABLE FUNCTIONS

Date/Time run: 01/04/80 17:54:26   Temperature: 27.0

Time

Figure 12-3

```
.TRAN .05 1 0 .01
.PROBE
.END
```

The output of this analysis is shown in Figure 12-3.

12.2.3. Voltage and Current Sources Using Laplace Transforms

A transfer function in the s-plane can be calculated using either the E device or the G device and the Laplace transform of the function. This source requires an input voltage. The output of the device is the Laplace transform of the input voltage as either a voltage or a current. The usual rules for the use of the Laplace transform apply. The general form of the sources for Laplace use is

```
E<name>  <+node>  <-node>  LAPLACE  {expression}  =
+                                   {transform}
G<name>  <+node>  <-node>  LAPLACE  {expression}  =
+                                   {transform}
```

Once again, both {expression} and {transform} must each fit on one and only one line. An example of both of the sources is

```
ELAPLACE 9 0 LAPLACE {V(1)} = {(1E4/(PWR(S,2)+40*S+1E4))}
GLAPLACE 2 0 LAPLACE {V(5)} = {(20/(S*(.02*S+1)))}
```

The Laplace variable, s, is a complex frequency variable. For this reason it is not possible to have voltage, current, or time in the transfer function. Input to the sources can be any mixture of voltages and currents that fit on one line. The keyword LAPLACE must be followed by a space. You can use both frequency and transient analysis with the LAPLACE sources.

As an example of the Laplace transform capability of PSpice, let us analyze the following system. The block diagram for the system is shown in Figure 12-4. The system is a proportional and integral control system for which the transfer function has been calculated to be

$$\frac{C(s)}{R(s)} = \frac{55.5s + 130}{s^3 + 12s^2 + 64.5s + 130}$$

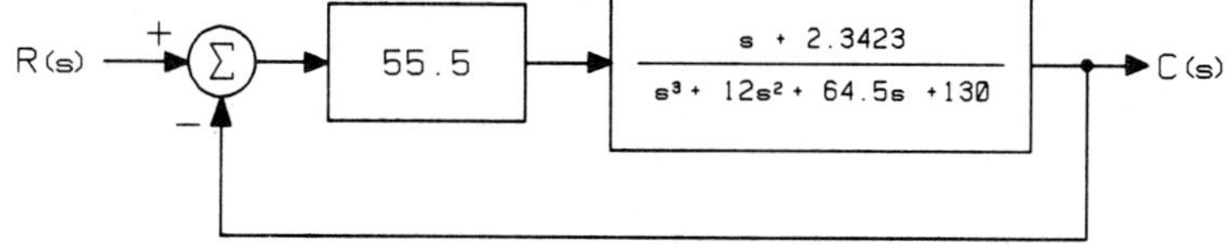

Figure 12-4

The system is designed to settle within 1.5 s and to have a maximum overshoot of less than 1.2 times the set point. The netlist for this analysis is

```
*LAPLACE DEMO
VIN 10 0 PWL(0,0 1U,1 4.5,1)
RIN 10 0 1E6
ERC 5 0 LAPLACE {V(10)} =
+    {(55.5*S+130)/(PWR(S,3)+12*PWR(S,2)+64.5*S+130)}
RRC 5 0 1E6
.TRAN .05 4 0 0.02
.PROBE
.END
```

The output of the analysis is shown in Figure 12-5. You can see that the system meets the required specifications.

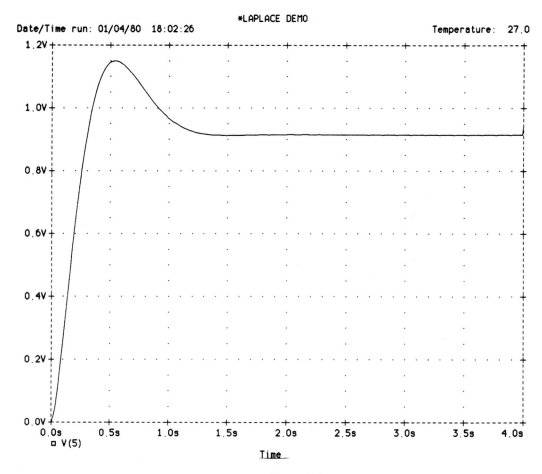

Figure 12-5

SUMMARY

In this chapter have been presented some additional capabilities of PSpice. The Monte Carlo analysis and the analog behavioral modeling capabilities of PSpice have been briefly explored. This is by no means all that PSpice can do. PSpice also has a digital option that has not been explored at all in this book.

A

Real and Imaginary Impedances

To understand how the impedance functions are determined by PSpice, a brief review of the use of imaginary numbers is in order. We shall use a simple *RC* low-pass filter for this explanation. The filter is shown in Figure A-1. The netlist to analyze this filter is

```
*SIMPLE RC LOW PASS FILTER DEMO
VIN 1 0 AC 5
R1 1 2 1K
C1 2 0 .16U
.AC DEC 25 10 1E5
.PROBE
.END
```

Note that in this circuit the ac input voltage is 5 V peak instead of the usual 1 V peak. This is done to force the use of the voltage function for calculating impedance, instead of just letting the voltage be 1 V peak. We shall analyze the circuit at a frequency of 500 Hz for both the real and the imaginary impedance. Shown in Figure A-2 is the usual complex impedance plane for our reference. The magnitude of the impedance of the circuit is 2236 Ω, at an angle of −63.4°. The magnitude of the applied voltage is 5 V at an angle of 0°. The magnitude of the current in the circuit is simply

$$|I| = \frac{|E|}{Z} = \frac{5}{2236} = 0.002236 \text{ A}$$

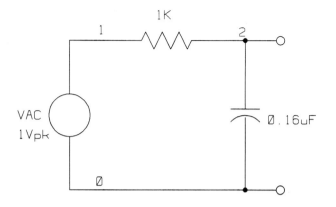

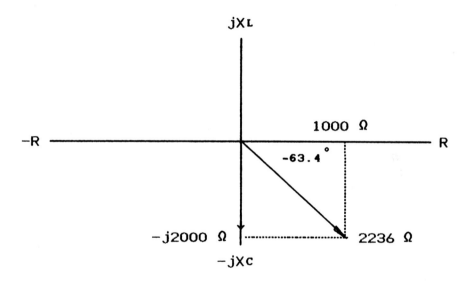

Figure A-1

Figure A-2

It is to be remembered that the current in a capacitor leads the voltage by 90°. Thus the phase angle of the current for this circuit at this frequency is +63.4°. Since the current is found using the magnitude of the impedance, it also is a magnitude that is made up of a real and an imaginary component. To find the real and imaginary components of the current we use Euler's equation:

$$I = 0.002236(\cos 63.4° + j \sin 63.4°)$$
$$= 0.001 + j0.002 \text{ A}$$

We see that the capacitive current is 2 mA and the resistor current is 1 mA. The magnitude of the voltage is

$$|E| = 5 \text{ V}$$

The impedance of the circuit can be found by using Ohm's law:

$$|Z| = \frac{|E|}{|I|} = \frac{|E|}{_RI\text{-} +_cjl} = \frac{5}{0.001 + j0.002}$$

After rationalizing the denominator,

$$|Z| = \frac{5(0.001 - j0.002)}{0.001^2 + 0.002^2}$$

The numerator now contains an expression for the power in the circuit. The denominator contains the sum of the squares of both the resistor and capacitor current. This is the square of the magnitude of the current in the circuit. We shall use this to find the real and imaginary components of the input impedance of the circuit.

To find the real part of the input impedance we must find the real part of the power of the circuit. This is

$$P = V_{in} \, \text{Re}(I_{in})$$

and the real part of the input impedance of the circuit is determined to be

$$\text{Re}(Z_{in}) = \frac{V_{in} \, \text{Re}(I_{in})}{|I^2|} = \frac{5 \times 0.001}{(0.002236)^2} = 1000 \; \Omega$$

and this is correct for this circuit. The imaginary component of the impedance is found in a similar manner using the imaginary component of the current in the numerator of the equation.

$$\text{Im}(Z_{in}) = \frac{V_{in} \, \text{Im}(I_{in})}{|I^2|} = \frac{5 \times 0.002}{(0.002236)^2} = 2000 \; \Omega$$

which we also know to be correct from Figure A1. The output impedance is done in a similar manner but using the real and imaginary components of the voltage developed across the output by a current generator.

B

PSpice Commands

Following is a short compendium of the commands and statements available in PSpice. Commands discussed in the text are not elaborated on here.

B.1. DEVICE STRUCTURES

* A comment line that starts in the first column of the netlist
; A comment line that starts after the last part of a command or device line.
B GaAsFET

```
B<name> <drain> <gate> <source> <model name> [AREA]
```

Parameter		Unit	Default
LEVEL	Model type 1 or 2		1
VTO	Pinch-off voltage	volt	−2.5
ALPHA	Saturation voltage	volt^{-1}	2.0
BETA	Transconductance coefficient	ampere/volt^2	0.1
B	Doping tail extending parameter (level = 2 only)	volt	0.3
LAMBDA	Channel-length modulation	volt^{-1}	0
TAU	Conduction current delay time	second	0
RG	Gate ohmic resistance	ohm	0

Parameter		Unit	Default
RD	Drain ohmic resistance	ohm	0
RS	Source ohmic resistance	ohm	0
IS	Gate *p-n* saturation current	ampere	1e–14
N	Gate *p-n* emission coefficient		1
M	Gate *p-n* grading coefficient		0.5
VBI	Gate *p-n* potential	volt	1.0
CGD	Zero-bias gate–drain capacitance	farad	0
CGS	Zero-bias gate–source capacitance	farad	0
CDS	Drain–source capacitance	farad	0
FC	Forward-bias depletion capacitance coefficient		0.5
VTOTC	VTO temperature coefficient	volt/°C	0
BETATCE	Beta exponential temperature coefficient	%/°C	0
KF	Flicker noise coefficient		0
AF	Flicker noise exponent		1

C Capacitor

```
        C<name> <+node> <-node> <model name> <value> [IC]
```

D Diode

```
        D<name> <anode> <cathode> <model name> [AREA]
```

E Voltage-controlled voltage source (VCVS)

```
        E<name> <output nodes> <input nodes>
        +       <control parameters> <gain factor>

        E<name> <output nodes> POLY(x) <input nodes>
        +       <polynomial coefficients>
```

F Current-controlled current source (CCCS)

```
      F<name> <output nodes> <controlling V device>
      +       <gain factor>

      F<name> <output nodes> POLY(x) <controlling V device>
      +       <polynomial coefficients>
```

G Voltage-controlled current source (VCCS)

```
      G<name> <output nodes> <input nodes> <transconductance>

      G<name> <output nodes> POLY(x) <input nodes>
      +       <polynomial coefficients>
```

H Current-controlled voltage source (CCVS)

```
H<name> <output nodes> <controlling V device>
+        <tranresistance>

H<name> <output nodes> POLY(x) <controlling V device>
+        <polynomial coefficients>
```

I Current source, independent

```
I<name> <+node> <-node> [DC] <value> [AC <magnitude>
+        <phase>] [<transient function>]
```

J Junction field-effect transistor

```
J<name> <drain> <gate> <source> <model name>
+        [Area parameter>]
```

K Inductor coupling coefficient

```
K<name> L<inductor name> L<inductor name> * * *
+        <coupling factor>

K<name> L<inductor name> L<inductor name> * * *
+        <model name> [<size>]
```

L Inductor

```
L<name> <+node> <-node> [model name] <value>
+        [IC = <initial value>]
```

M MOSFET

```
M<name> <drain> <gate> <source> <substrate> <model name>
+        [L = <value>] [W = <value>] [AD = <value>]
+        [AS = <value>] [PS = <value>] [NRD = <value>]
+        [NRS = <value>] [NRG = <value>] [NRB = <value>]
```

Parameter		Units	Defaults
LEVEL	Model type: 1, 2, 3		1
L	Channel width	meter	DEFL
W	Channel width	meter	DEFW
LD	Lateral diffusion (length)	meter	0
WD	Lateral diffusion (width)	meter	0
VTO	Zero-bias threshold voltage	volt	0
KP	Transconductance	ampere/volt^2	$2e-5$
GAMMA	Bulk threshold parameter	volt$^{1/2}$	0

Parameter		Units	Defaults
PHI	Surface potential	volt^{-1}	0.6
LAMBDA	Channel-length modulation (level = 1 or 2)	volt^{-1}	0
RD	Drain ohmic resistance	ohm	0
RS	Source ohmic resistance	ohm	0
RG	Gate ohmic resistance	ohm	0
RB	Bulk ohmic resistance	ohm	0
RDS	Drain–source shunt resistance	ohm	infinite
RSH	Drain–source diffusion sheet resistance	ohm/sq.	0
IS	Bulk *p-n* saturation current	ampere	1e − 14
JS	Bulk *p-n* saturation current/area	ampere/meter^2	0
PB	Bulk *p-n* potential	volt	0.8
CBD	Bulk–drain zero-bias *p-n* capacitance	farad	0
CBS	Bulk–source zero-bias *p-n* capacitance	farad	0
CJ	Bulk *p-n* zero-bias bottom capacitance/area	farad	0
CJSW	Bulk *p-n* perimeter capacitance/length	farad	0
MJ	Bulk *p-n* bottom gradine coefficient		0.5
MJSW	Bulk *p-n* sidewall grading coefficient		0.33
FC	Bulk *p-n* forward-bias capacitance coefficient		0.5
CGSO	Gate–source overlap capacitance/channel width	farad/meter	0
CGDO	Gate–drain overlap capacitance/channel width	farad/meter	0
CGBO	Gate–bulk overlap capacitance/channel length	farad/meter	0
NSUB	Substrate doping density	1/cm^3	0
NSS	Surface state density	1/cm^2	0
NFS	Fast surface state density	1/cm^2	0
TOX	Oxide thickness	meter	infinite
TPG	Gate material type: +1 Opposite of substrate −1 Same as substrate −0 Aluminum		+1
XJ	Metallurgical junction depth	meter	0
UO	Surface mobility	cm^2/volt·second	600
UCRIT	Mobility degradation critical field (level 2)	volt/cm	1E4
UEXP	Mobility degradation exponent (level 2)		0
UTRA	Mobility degradation transverse field coefficient	not used	
VMAX	Maximum drift velocity	meter/second	0
NEFF	Channel charge coefficient (level 2)		1
XQC	Fraction of channel charge caused by drain		1
DELTA	Width effect on threshold		0
THETA	Mobility modulation (level 3)	volt^{-1}	0
ETA	Static feedback (level 3)		0
KAPPA	Saturation field factor (level 3)		0.2
KF	Flicker noise coefficient		0
AF	Flicker noise exponent		1

Q Bipolar junction transistor

```
Q<name> <collector> <base> <emitter> [<substrate>]
+       <model name> [area]
```

R Resistor

```
R<name> <+ node> <- node> [<model name>] <value>
```

S Voltage-controlled switch

```
S<name> <+ node> <- node> <+ controlling node>
+       <- controlling node> <model name>
```

Parameter		Unit	Default
RON	"On" resistance value	ohm	1
ROFF	"Off" resistance value	ohm	1E+6
VON	Control voltage of "on" state	volt	1
VOFF	Control voltage of "off" state	volt	0

This device acts as a voltage-controlled resistor. The resistance between the switch nodes depends on the voltage between the controlling nodes. RON and ROFF must be greater than 1/GMIN. This value is changeable using the .OPTIONS command. 1/GMIN is connected between the controlling nodes to keep them from floating. The ratio of ROFF to RON should not be greater than 1E+12.

T Transmission line

```
T<name> <+ node 1> <-node 1> <+ node 2> <- node 2>
+       ZO = <value> [TD = <value>] F = <value>
+       [NL = <value>]
```

V Independent voltage source

```
V<name> <+ node> <- node> [DC] <value> [AC <magnitude>
+       <phase>] [<transient function>]
```

W Current-controlled switch

```
W<name> <+ node> <- node> <+ controlling node>
+       <-controlling V device> <model name>
```

Parameter		Unit	Default
RON	"On" resistance value	ohm	1
ROFF	"Off" resistance value	ohm	1E+6
VON	Control current of "on" state	ampere	0.001
VOFF	Control current of "off" state	ampere	0

This device acts as a voltage-controlled resistor. The resistance between the switch nodes depends on the current through the controlling V device. RON and ROFF must be greater than 1/GMIN. This value is changeable using the .OPTIONS command. 1/GMIN is connected between the controlling nodes to keep them from floating. The ratio of ROFF to RON should not be greater than 1E+12.

X Subcircuit notation

```
X<name> <node> * * * <subcircuit name>
```

B.2. COMMAND STRUCTURES

.AC   Ac sweep command

```
.AC <sweep type> <points value> <start frequency>
         <end frequency>
```

.DC   Dc sweep command

```
.DC <sweep type> <swept device> <start value>
+              <end value> <increment value>
```

.END   End of netlist

```
.END
```

.ENDS   End of subcircuit netlist

```
.ENDS [subcircuit name]
```

.FOUR   Fourier analysis

```
.FOUR <frequency of analysis> <output variable> * * *
```

.IC   Initial condition for transient analysis

```
.IC V<node> = <value> * * *
```

.INC   Include file

```
.INC [file name]
```

This command is used to insert a file outside the present netlist into the netlist being analyzed. Only four levels of including may be used. All parameters in the included file are read into the analysis and take up space in RAM whether or not they are needed in the analysis.

.LIB   Library file

```
.LIB [library file name]
```

.MC   Monte Carlo analysis

```
.MC <num runs> [DC] [AC] [TRAN] <output variable> YMAX
+    [LIST] [OUTPUT <output required>]
```

.MODEL   Model specification for devices

```
.MODEL <name> <model type> [<parameter name> = <value>
+       [tolerance]] * * *
```

.NODESET   Preset node voltage or current

```
.NODESET <V<node> = <value>> * * *
```

.NOISE   Noise analysis

```
.NOISE V<nodes> <name> [internal value]
```

.OP   Operating bias point analysis

```
.OP
```

.OPTIONS   Set program default options

```
.OPTIONS [option name] * * * [<option name> = <value>]
```

Option	Definition
ACCT	Summary and time accounting information is output at the end of all analyses
LIST	List circuit elements and devices
NODE	Node table of netlist is output

Option	Definition
NOECHO	Stops listing of the input netlist
NOMOD	Stops listing of model parameters and updated temperature values
NOPAGE	Stops paging and printing of the banner for each major section of output
OPTS	Values of all options are output
WIDTH	Changes width of output file

Option	Definition	Default	Unit
ABSTOL	Accuracy of currents	1 Pa	ampere
CHGTOL	Accuracy of charges	0.01 pC	coulomb
CPTIME	CPU time of the run	1E6	second
DEFAD	MOSFET default drain area (AD)	0	meter^2
DEFAS	MOSFET default source area (AS)	0	meter^2
DEFL	MOSFET default length (L)	100 μm	meter
DEFW	MOSFET default width (W)	100 μm	meter
GMIN	Minimum conductance of any branch	1E−12	siemens
ITL1	Dc and bias-point "blind" iteration limit	40	
ITL2	Dc and bias-point "best guess" iteration limit	20	
ITL4	Iteration limit for any point in transient analysis	10	
ITL5	Total iteration limit of all points in transient analysis (ITL5 = 0 calculates an infinite number of points)	5000	
LIMPTS	Maximum number of points allowed for any print or plot table	201	
NUMDGT	Number of digits in print tables (eight maximum)	4	
PIVREL	Relative magnitude required for pivot point in solving matrix	0.001	
PIVTOL	Absolute magnitude required for pivot in solving matrix	1E−13	
RELTOL	Relative accuracy of voltages and currents	0.001	
TNOM	Default temperature	27	°C
TRTOL	Transient analysis accuracy	7	
VNTOL	Accuracy of voltages	1E−6	

.PLOT   Text graphics plot of parameters

```
.PLOT [DC] [AC] [NOISE] [TRAN] <output variable> * * *
+      <min range> <max range>
```

.PRINT   Print parameters to a table

```
.PRINT [DC] [AC] [NOISE] [TRAN] <output variable> * * *
```

.PROBE   High-resolution graphics postprocessor

```
.PROBE [<output variable> * * *]
```

.SENS   Sensitivity of components to voltages and currents

```
.SENS <output variable> * * *
```

Dc sensitivity of components to variations in component values and model parameters is output. If current sensitivity is needed, it must be measured through a voltage source. Large quantities of data can be generated by this command.

.SUBCKT   Subcircuit call

```
.SUBCKT <name> <nodes> * * *
```

.TEMP   Temperature value

```
.TEMP <temperature value> * * *
```

.TF   Transfer function of circuit

```
.TF <output variable> <input source>
```

.TRAN   Transient analysis specification

```
.TRAN[/OP] <print time steps> <final time value>
+         [<no print time>] [<maximum time step>]
+         [UIC]
```

.WIDTH   Width of output file

```
.WIDTH OUT <value>
```

This command sets the width of the output file. The value is in columns of output. It must be either 80, which is the default value, or 132, which is the maximum value.

Bibliography

ANTOGNETTI, PAOLO, AND MASSOBRIO, GIUSEPPE, *Semiconductor Device Modeling with SPICE.* New York: McGraw-Hill, 1988.

BANZHAF, WALTER, *Computer-Aided Circuit Analysis Using SPICE.* Englewood Cliffs, N.J.: Prentice Hall, 1988.

BOYLE, G., COHN, D., PEDERSON, D., AND SOLOMON, J., Macromodeling of integrated circuit operational amplifiers, *IEEE Journal of Solid-State Circuits,* Vol. SC-9, No. 6, December 1974, pp. 353–364.

BOYLESTAD, ROBERT, AND NASHELSKY, LOUIS, *Electronic Devices and Circuit Theory.* Englewood Cliffs, N.J.: Prentice-Hall, 1987.

BUCEK, VICTOR J., *Control Systems.* Englewood Cliffs, N.J.: Prentice Hall, 1989.

CHEN, WAI-KAI, *Passive and Active Filters: Theory and Implementations.* New York: Wiley, 1986.

EBERS, J. J., AND MOLL, J. L., Large signal behaviour of junction transistors, *Proceedings IRE,* Vol. 42, December 1954, pp. 1761–1772.

HUELSMAN, L. P., AND ALLEN, P. E., *Introduction to the Theory and Design of Active Filters.* New York: McGraw-Hill, 1980.

HUGHES, FREDERICK W., *Op-Amp Handbook.* Englewood Cliffs, N.J.: Prentice Hall, 1986.

JILES, D. C., AND ATHERTON, D. L., Theory of ferromagnetic hysteresis, *Journal of Magnetism and Magnetic Materials,* Vol. 61, 1986, p. 48.

MEARES, LAWRENCE G., AND HYMOWITZ, CHARLES E., *Simulating with Spice.* San Pedro, Calif.: Intusoft, 1988.

MICROSIM CORPORATION, *PSpice Manual.* Irvine, Calif.: MicroSim Corporation, 1989.

336

NATIONAL SEMICONDUCTOR, *Linear Databook*. Santa Clara, Calif.: National Semiconductor Corporation, 1978.

RASHID, MUHAMMAD H., *SPICE for Circuits and Electronics Using PSpice*. Englewood Cliffs, N.J.: Prentice Hall, 1990.

SCHICHMAN, H., AND HODGES, D. A., Modeling and simulation of insulated gate field effect transistor switching circuits, *IEEE Journal of Solid-State Circuits,* Vol. SC-3, September 1968, pp. 285–289.

TEXAS INSTRUMENTS, *Operational Amplifier Macromodels: Linear Circuits Manual*. Dallas, Tex.: Texas Instruments, Inc., 1990.

TUINENGA, PAUL W., *SPICE: A Guide to Circuit Simulation and Analysis Using PSpice*. Englewood Cliffs, N.J.: Prentice Hall, 1988.

YOUNG, THOMAS, *Linear Integrated Circuits*. New York: Wiley, 1981.

Index